STRIKE FROM THE SEA

U.S. Navy Attack Aircraft from Skyraider to Super Hornet 1948–Present

Tommy H. Thomason

specialtypress
PUBLISHERS AND WHOLESALERS

Specialty Press
39966 Grand Avenue
North Branch, MN 55056
Phone: 651-277-1400 or 800-895-4585
Fax: 651-277-1203
www.specialtypress.com

Edit by Mike Machat
Layout by Chris Fayers

ISBN 978-1-58007-132-1
Item No. SP132

Library of Congress Cataloging-in-Publication Data

Thomason, Tommy H.
   Strike from the sea : U.S. Navy Attack Aircraft from Skyraider to Super Hornet 1948–Present / by Tommy H. Thomason.
      p. cm.
   ISBN 978-1-58007-132-1
   1. Attack planes—United States—History. 2. Bombers—United States—History. 3. United States. Navy—Aviation—History. 4. Aeronautics, Military—United States—History. I. Title.
   UG1242.A28T48 2009
   358.4'2830973—dc22
                                    2008056146

Printed in China
10 9 8 7 6 5 4 3 2 1

On the Front Cover:
Serving as the backbone of U.S. Navy carrier-based aircraft operations today, the F/A-18A/C Hornets and the F/A-18E/F Super Hornets fly in a wide variety of roles from air-superiority and ground-attack missions. In 2009, these will be joined by the EA-18G providing electronic reconnaissance and attack. Here, a VFA-147 Argonauts F/A-18C, armed with AIM-9 Sidewinders and AGM-65 Mavericks, is in low holding above *Nimitz* in September 1996 awaiting approval to enter the pattern for landing. The tailhook has already been lowered en route to the carrier to ensure that it is available. (*U.S. Navy photo DNSC-04-08802*)

On the Back Cover (top left):
The North American AJ Savage was designed and built to carry a single Mark III atom bomb, and was the largest aircraft ever deployed on an aircraft carrier until the jet-powered Douglas A3D Skywarrior became operational in 1956. Here, an AJ-1 executes a wave-off and go-around while operating from the USS *Coral Sea* (CVB-43) on 31 August 1950. (*National Archives*)

On the Back Cover (top right):
The Douglas A4D Skyhawk was the longest continuous production tactical aircraft in the Navy inventory, with nearly 3,000 built between 1954 and 1979. The multiple bomb rack adapters shown here allowed the A4D to be more productive when flying conventional weapons-delivery missions, as a greater number of smaller bombs provided more close-air-support capability than only two or three large ones. (*U.S. Navy via author's collection*)

On the Back Cover (bottom):
The Grumman EA-6B Prowler, shown flying over the USS *John F. Kennedy* (CVA-67), added two more Naval Flight Officers to its crew to operate the aircraft's advanced ECM equipment. The EA-6B's radar-jamming capability could be tailored to specific missions by the mix of AN/ALQ-99 jamming pods carried, while the coated canopies of the EA-6B shown here protected the crew from high-power transmissions emitted by its jamming pods. The four-place Prowlers along with the E-2C Hawkeye are the last Grumman-built airplanes to serve with carrier air wings. (*U.S. Navy photo*)

On the Title Page:
Although the Grumman A-6 Intruder obviously couldn't carry all of these weapons at one time, it still made an impressive publicity photo to show the wide array of ordnance the airplane carried on a variety of attack missions. These included Mk 81 250-pound bombs, Mk 82 500-pound bombs, Mk 84 2,000-pound bombs, Zuni rockets, and AGM-12A Bullpup air-to-ground missiles, among others. Stores shown draped in front of the airplane represent classified nuclear weapons. The aircraft itself carries five 300-gallon fuel tanks, including one not visible on the centerline. (*Grumman History Center*)

Distributed in the UK and Europe by
Crécy Publishing Ltd
1a Ringway Trading Estate
Shadowmoss Road
Manchester M22 5LH England
Tel: 44 161 499 0024
Fax: 44 161 499 0298
www.crecy.co.uk
enquiries@crecy.co.uk

TABLE OF CONTENTS

DEDICATION

To Navy carrier-based attack pilots

As the fictional Rear Admiral George Tarrant said in James A. Michener's Korean War novel, *The Bridges at Toko-Ri*:
"Where do we get such men? They leave this ship and they do their job. Then they must find this speck lost somewhere on the sea.
When they find it, they have to land on its pitching deck. Where do we get such men?"

PREFACE

"**Y**ou can't tell the players without a score card!" screamed the vendors at baseball parks in times past. This is particularly true when the players' names have changed, sometimes more than once, in the past 60 years. In this book, I use the name being used at the time. The following summarizes the changes. Also provided, at the risk of oversimplifying, is a summary of the process by which aircraft programs are approved and funded.

Navy Airplane Contractors

In 1948, there were many more separate airplane manufacturing entities than there are today. Many were named for their founders. Navy prime contractors (suppliers of complete airplanes) included Boeing, Convair (Consolidated Vultee), Douglas, Grumman, Lockheed, Martin, McDonnell, North American, and Vought. When the dust from mergers and acquisitions had finally settled in the late 1990s, two-thirds were gone or no longer independent. The surviving entities were Boeing, Lockheed-Martin, and Northrop Grumman.

Consolidated Aircraft was founded by Reuben H. Fleet in Buffalo, New York, in 1923. Vultee Aircraft was created in 1939, although it had existed in other forms beginning in 1932 with Gerard Vultee as a founder. Consolidated and Vultee merged and became Consolidated Vultee in 1943. Later, the entity became known as Convair. It was acquired by General Dynamics[1] in 1953, becoming the Convair division, with major facilities in San Diego, California, and Fort

Worth, Texas. General Dynamics divested itself of many operations in the mid 1990s, with the plant in Fort Worth that had produced the B-36, B-58, F-111, and F-16 being acquired by Lockheed, along with the F-16 program, in 1993.

Grumman was founded in 1929 by Leroy Grumman and located on Long Island, New York. It fell on hard times in the early 1990s and was acquired by Northrop in 1994.[2] The resulting enterprise was named Northrop Grumman to capitalize on Grumman's reputation with the Navy.

Vought was originally Lewis and Vought, with cofounder Chance Vought renaming the company in 1922 as the Chance Vought Corporation. As of 2008, it is still a separate, albeit wholly owned, entity but no longer a designer and builder of complete aircraft. By the time its A-7 Corsair II had been retired, Vought had had various owners and names and become a subcontractor for airframe structure to prime contractors. It was renamed Vought Aircraft Industries in 2000 after its purchase from Northrop Grumman by the Carlyle Group.

North American Aviation (established as a manufacturing company in 1934 with James "Dutch" Kindelberger as its president) and Douglas (founded by Donald W. Douglas in 1921) were also absorbed by other, more successful companies, eventually winding up as part of Boeing. North American was merged with Rockwell Standard in 1967, the corporation becoming North American Rockwell. In 1996, when known as Rockwell International, the parent company was bought by Boeing. Douglas was acquired by McDonnell Aircraft Corporation (founded by James S. McDonnell in

1939) in 1967, with the parent company being named McDonnell Douglas. McDonnell Douglas merged with Boeing in 1997, but the result was named The Boeing Company. McDonnell Douglas was identified as a separate entity for a while. Now its facilities are an integral part of Boeing Integrated Defense Systems.

The Glenn L. Martin Company was incorporated in Los Angeles, California, in 1912. Donald Douglas, James McDonnell, Chance Vought, Dutch Kindelberger, and Larry Bell all worked for the company at various times. The Loughead brothers established the Loughead Aircraft Manufacturing Company in Santa Barbara, California, in 1916. A few years after his original company failed, Loughead established the similar-sounding Lockheed Aircraft Company in Hollywood, California, in 1926. Lockheed and Martin merged in 1995.

Attack Airplane Designation Summary

In 1948, the newly independent U.S. Air Force changed its designation system for aircraft. P for Pursuit was replaced by F for Fighter. (F had been previously assigned to photoreconnaissance airplanes). The dash number of a fighter type stayed the same when the letter change was made.

A for Attack was dropped entirely by the Air Force. The designation had been used since 1926 for close air support and light bombardment aircraft types, with the sequential assignment of numbers having reached 45. B for Bomber was retained, and all the attack types still in the inventory and those in development received bomber designations, with the exception of the Curtiss XA-43 project, which was canceled, and a handful of single-engine Douglas A-24s (initially developed by the Navy as the SBD), which became F-24s. The sole type operational, the North American A-26 light bomber, became the B-26, previously the designation of the Martin Marauder. The Douglas XA-42 became the XB-42. On the other hand, the Convair XA-44 became the XB-53 (none built) and the Martin XA-45 became the XB-51.

The deletion of the A for Attack designation didn't mean that the Air Force no longer had an attack-type mission. It was now collateral duty for fighter types or, in a few instances, bomber types.

The Navy designation prior to 1962 is described in Chapter Two. Ironically, the Navy adopted A for Attack in place of B for Bomber the same year that the Air Force decided to eliminate it.

The change to a common designation system for all the services dictated by the Department of Defense in 1962 retained the Attack designation, restarting the number sequence at one. The new attack designations mainly affected a handful of Navy aircraft that were still in service or in development:

Aircraft	Old	New
Douglas Skyraider	AD	A-1
North American Savage	AJ	A-2
Douglas Skywarrior	A3D	A-3[3]
Douglas Skyhawk	A4D	A-4
North American Vigilante	A3J	A-5
Grumman Intruder	A2F	A-6

Subsequent attack aircraft, which included new Air Force developments, were designated in sequence:

- A-7: Vought Corsair II single-engine jet for Navy and then the Air Force
- AV-8: McDonnell license-built, VTOL-capable Hawker Siddeley Harrier
- A-9: Northrop twin-engine, single-seat attack airplane that competed for an Air Force close air support requirement and lost
- A-10: Fairchild Republic Thunderbolt II twin-engine, single-seat attack airplane that competed with the A-9 and won
- A-11: The designation given to the Lockheed YF-12 Mach 3 fighter when it was first revealed in February 1964. The announcement provided cover for the similar and still secret Lockheed A-12 reconnaissance airplane being developed for the Central Intelligence Agency to replace the U-2. Within Lockheed, the A stood for Archangel, with the A-12 being the twelfth in a series of design studies for the mission.
- A-12: General Dynamics Avenger II attack airplane. There was no risk of confusion with the CIA's A-12 since it had been retired 20 years earlier.

The next attack airplane designation to be used was A-18. The skip of five numbers resulted because it was to be a unique attack configuration of the F-18. When the configurations merged, the F-18 and A-18 became the F/A-18. Designation purists were offended, but there was worse to come with the F-35.

Even before the sequential system of designation of fighter and attack aircraft was discarded in favor of a pragmatic and ad hoc approach, there were designation oddities. Ironically, the Douglas B-26 once again became an A-26 when the B-26K was developed for counter-insurgency missions. Since a treaty between the U.S. and Thailand forbade the basing of bombers in Thailand, it became the A-26A. Another adaptation of a designation number resulted in the A-37; it was an armed close air support derivative of the T-37, a two-seat primary jet trainer.

In any event, the attack designation again appears to be defunct, this time in both the Air Force and the Navy. In fact, the new F-35 (another ad hoc designation stemming from its origination as the X-35) is a strike fighter like the F/A-18, but has not been burdened with an A for Attack or even a V for Vertical capability in the case of the Marine Corps/Royal Navy STOVL variant, F-35B.

The Procurement Process

There are many participants in the airplane and armament procurement process who represent the customer. The process itself has not fundamentally changed in the past 60 years, in part because it is defined by law, but the relative influence and activism of some of the players, particularly in the civilian chain of command, has varied.

The deployable Naval aviators are the most dependent but probably the least influential on what is eventually issued to them. Their wing commanders and fleet admirals are somewhat more effective in changing the status quo but sometimes disagree. Since the most important function of the air wing is to protect the carrier, the fighter community receives a high priority for improvements and replacements for their airplanes. The attack community was at somewhat of a disadvantage because it was split into two camps, heavy and light, with differing views on the respective value provided and investment required for deep strike versus close air support, to name two important attack missions. Further, the deployed Navy is divided into East and West Coast organizations and further segmented into fleets, with differing operational requirements based on the geopolitical peculiarities of the respective fleet assignment.

All these constituents report up to the Chief of Naval Operations (CNO), who has a staff responsible for prioritizing and synthesizing the fleets' requirements into a plan. For convenience, I'll refer to it as OpNav. A procurement organization reporting to the CNO deals with developing and buying airplanes in accordance with those requirements and plans. From 1921 through 1960, it was known as the Bureau of Aeronautics (BuAer). At that time it was combined with the Bureau of Ordnance to become the Bureau of Naval Weapons (BuWeps). In 1966, the Navy reorganized again and aircraft development and procurement was the responsibility of the Naval Air Systems Command (NavAir). By whatever name, one of its major tasks is to contract for and manage new and upgraded airplane programs. Sometimes a formal competition is used to select the contractor for a program and sometimes a decision is made to upgrade an existing airplane.

As with all the U.S. services, the senior official in Navy is a civilian Secretary, referred to as SecNav. This is a political appointment, with selectees having varying levels of experience. Francis P. Matthews, a lawyer from Omaha who became SecNav in 1949, admitted to not having much of a nautical background. He famously said, "I do have a rowboat at my summer home."[4] The other extreme was John F. Lehman Jr., who was an A-6 bombardier/navigator (B/N) in the Navy Reserves when he became Secretary of the Navy in 1981.

The service secretaries report to the Secretary of Defense (SecDef), who is a member of the President's cabinet and another political appointment. This position was created in 1947 with James V. Forrestal as the first SecDef. Until Robert S. McNamara was appointed in 1961, the Office of the Secretary of Defense (OSD) did not usually delve too deeply into the inner workings and decisions of each service, primarily adjudicating disputes and allocating budgets between services. McNamara not only instituted systems analysis as the basis for making decisions, he required that those decisions and the supporting analysis for major programs be reviewed in detail by OSD before receiving approval. In the case of the TFX program, which resulted in the F-111 for both the Navy and the Air Force, he reversed the source selection from Boeing to General Dynamics.

The process does not end there, however. The President has to submit the Secretary of Defense's budget request to Congress each year for authorization and funding for the upcoming fiscal year. Congress will then modify the request as it sees fit. Its decision is literally law. No program can be initiated or continue without such authorization and also being funded with an appropriation. There are four separate Armed Services committees in the House and Senate, one in each chamber for authorization and one for appropriation. The authorization and appropriations bills from the committees have to be approved by the full House and Senate and then reconciled by the two bodies before becoming law. This is a prolonged and messy process, with congressmen and senators having to take into consideration the views of all their constituents and those of their staffs. They also solicit advice from active duty and retired military personnel and civil servants, both formally in hearings and informally. The airplane contractors all maintain offices in the Washington, D.C., area that have the responsibility of influencing Congress with respect to their programs, not to mention those of their competitors.

The Navy has a process for testing and evaluating systems against contract and operational requirements before accepting them. The Navy's Board of Inspection and Survey (BIS) oversees this. Most of the testing is done at Naval Air Station (NAS) Patuxent River, Maryland, the home of the Naval Air Test Center, including shore-based catapult takeoff and arresting landing trials. Qualifications at sea are accomplished on whatever carrier can be made available in the size class required. Nuclear weapons clearance is done at the Naval Weapons Evaluation Facility, which is collocated with the Air Force at Kirkland Air Force Base, Albuquerque, New Mexico. Missile trials are usually accomplished at NAS Point Mugu, California.

One relatively little-known Navy organization requires special mention, however, and that is the Naval Air Weapons Station at China Lake, California. It is located about 40 miles north of Edwards Air Force Base and consists of 1.1 million acres that are uninhabited except for government facilities. It was originally established as the Naval Ordnance Test Station at Inyokern in 1943, evaluating and improving the British air-to-surface rocket and the U.S. Mk 13 aerial torpedo. The proximity-firing device for the first atomic bombs was also developed and qualified there. Since then it has produced many innovations and improvements in airborne weapons. Probably the best known product of its scientists and engineers is the heat-seeking Sidewinder missile, but its creations include the Mighty Mouse and Zuni folding fin rockets, the multiple-ejector rack, several "Eye" designated free-fall bombs, the AGM-45 Shrike antiradar missile, and the AGM-123 laser-guided missile. It is a very unusual integration of civil service and military personnel in a laboratory environment that combines basic research and practical engineering to produce innovative and practical solutions to operational problems and requirements.

ACKNOWLEDGMENTS

Hopefully, I will not fail to recognize any of the many people who made direct and significant contributions to the content of this book.

First and foremost, some aerospace companies not only value their heritage, but also freely provide access to it and support a cadre of volunteers who respond to questions and requests. I am particularly appreciative of the support provided to me by Dick Atkins and his coworkers at the Vought Aircraft Heritage Foundation; Larry Feliu and Tom Griffin at the Grumman Historical Center; and Stan Piet at the Glenn L. Martin Maryland Aviation Museum. Lockheed Martin doesn't have readily accessible archives, but they do have employees like Eric Hehs, Denny Lombard, and Kimberly Prato who are quick to respond to requests for information on more recent programs.

This book has also benefited immensely from efforts by the staff at the Naval Historical Center in Washington, D.C., specifically Joe Gordon and Mark Evans; the ladies in the Still Pictures Research Room at the National Archives in College Park, Maryland; and the staff at the Jay Miller Aviation History Collection in Little Rock, Arkansas. Not only have my visits always been very productive, they have been enjoyable.

Many individuals have made significant contributions of material to the book and/or commented knowledgeably on drafts. These include David Aiken, Kevin Austin, Ed Barthelmes, Steve Bradish, Tony Buttler, Tony Chong, John Cook, Will Dossel, Norm Filer, Walt Fink, Art Hanley, Joseph Hegedus, Bob Hickerson, Clarence Hutchison, Dennis Jenkins, Allison Jones, Craig Kaston, Mike Kimbell, Carroll Lefon, Mark Morgan, Rick Morgan, William Norton, Frank O'Brimski, William Paisley, Scott Pedersen, Boom Powell, Nicholas T. Spark, Barrett Tillman, David K. Stumpf, Nick Veronico, Hal Vincent, and Rod Wernicke.

Special thanks go to Steve Ginter, Bob Lawson, Jay Miller, Terry Panopalis, and Gary Verver for the many images they provided; Jim Rotromel for his patient assistance with weapons identification and description; Rosario (Skip) Rausa for editing; and Judith B. Currier for making the oral history of her father, George Spangenberg, available at http://www.georgespangenberg.com/.

Finally, many books written in the last few decades on U.S. Naval Aviation owe a debt of gratitude to the late Hal Andrews, this one even more than most. He preserved material and crafted well-written, meticulously researched articles for his and our enjoyment. He will always be remembered for his good nature, generosity, and scholarship.

INTRODUCTION

"Speak softly and carry a big stick." That's what Teddy Roosevelt said in April 1903. He sent 16 battleships and supporting ships around the world five years later, making 20 port calls on six continents. It became known as The Great White Fleet. The voyage was promoted as a diplomatic mission of peace. However, it also demonstrated that the U.S. Navy could operate globally in strength, capable of protecting America's interests abroad and the sea lanes of communication with America, as Roosevelt's big stick.

At the time, there were no aircraft carriers and few airplanes, the Wright brothers having first flown in December 1903. War at sea consisted of cannon-armed ships firing broadsides at each other using maneuver tactics that dated back to the early 17th century. The battleship, short for line-of-battle ship, was the largest, most heavily armored and armed. Its big guns were the main battery in confrontations of naval fleets. The outcome of actions was not necessarily determined by which side had the most or biggest battleships, however, since range, accuracy, and the rate of fire of the main battery were key determinants.

In 1921, General William L. "Billy" Mitchell demonstrated the vulnerability of ships to bombs dropped from aircraft by sinking the ex-German World War I battleship, *Ostfriesland.* Publicly and in most cases privately, the admirals were unimpressed. Nevertheless, construction of the first aircraft carrier, *Langley* (CV-1), as a conversion from the collier *Jupiter,* was already under way. The unprepossessing "Covered Wagon" was used to develop and evaluate operating techniques and tactics. Because of its humble origins, *Langley's* performance was unimpressive. It did not have the cruising speed to keep up with the fleet. Fortunately, the enhanced scouting and long-range strike functions that an aircraft carrier could provide were obvious.

Even more fortuitously, a ready source of real—in fact magnificent—aircraft carriers resulted from the 1921 naval limitations conference because the United States agreed to halt construction of four battle cruisers. Two of these were converted to aircraft carriers, *Lexington* (CV-2) and *Saratoga* (CV-3), bigger and faster than most of the battleships in service at the time. Commissioned in the late 1920s, these were followed by *Ranger* (CV-4), *Yorktown* (CV-10), *Enterprise* (CV-6), and *Wasp* (CV-7) during the next decade, the last smaller than its forebears to utilize the remainder of the carrier tonnage permitted by the Washington Treaty. The Navy also invested in the development of fighters and bombers specifically designed to operate from them. As a result, when World War II began the Navy had in place aircraft carriers, carrier-based aircraft, and people experienced in operating them.

The debacle at Pearl Harbor in December 1941 that found the U.S. Navy carriers at sea and out of harm's way only accelerated the transition of the main battery from big guns on battleships to airplanes operating from aircraft carriers. As at Agincourt in 1415, when the mounted French knights encased in armor were helpless against the arrows fired at long range from the longbows of the English archers, battleships were vulnerable to the torpedoes and bombs delivered by aircraft operating from carriers at far greater ranges than shells fired from ships could reach. By the end of World War II, the U.S. Navy had raised to a high level their proficiency at delivering devastation to ships and shore-based targets alike from carrier-based aircraft.

U.S. presidents since Harry Truman have asked "Where are the carriers?" in response to notification of a crisis developing somewhere in the world that affects U.S. interests. However, Truman didn't initially appreciate the cost versus the future benefit of a carrier battle group in the years immediately following the end of World War II. The U.S. and British fleets had sunk most of the ships of the opposing navies. Defense budgets were being slashed. Soon to be independent and demand at least a third of the military budget, the Army Air Forces was taking credit for ending the war with the atomic bomb and having the means to end any future conflict with this new and decisive weapon. If General Curtis LeMay and his supporters had their way, the Navy would have been relegated to a coast guard. The war with Japan had ended, but a new one with the Air Force had begun.

The Navy was faced with two challenges to its *raison d'être* and therefore funding from Congress. First, there was no longer any near-term concern about enemy ships. For all practical purposes, there weren't any challengers for control of the oceans and sea. Second, given that the delivery of the atomic bomb was all-important, even the physically smaller Mk 1 was far too heavy to be carried for any distance by existing carrier-based airplanes or even those in development.

The Navy concluded that to retain its fair share of the defense budget, it was not enough to protect the sea lines of communication vital to the economies of both the United States and its allies; provide a flexible and rapid military response to international crises; and defend the nation from sea-borne threats. It must have an atomic bomb delivery capability that did not require the use of foreign bases. The Navy immediately began development of bigger, longer-range carrier-based bombers, and was eventually successful in claiming a portion of the nuclear strike capability.

Also, as the invasion of South Korea by North Korea demonstrated, each of the nation's presidents in turn would come to value the deployed aircraft carrier's use as first responder to international crises with the ability to strike with conventional weapons. At the end of World War II, BuAer was in the process of completing the development of a new generation of carrier-based bombers and weapons for that purpose. The goal—to attack more effectively as well as increase the likelihood that the pilot would survive—remains the same today.

What follows is an account of the last 60 years of progress toward that goal, with both nuclear and conventional weapons.

USS **Ranger (CVA-61), *circa 1961.*** (U.S. Navy)

CHAPTER ONE

FORGED IN BATTLE

The SB2C Helldiver was the result of a 1938 competition to replace the Navy's SBD Dauntless dive-bomber. It suffered teething problems and was rejected by the captain of the first carrier that it was assigned to operationally in mid-1943. However, patience and persistence finally resulted in a modified design being acceptable for deployment in September 1943. (U.S. Navy photo via Robert L. Lawson collection)

Before World War II, the primary mission of the U.S. Navy's carriers was the destruction of an enemy fleet. Two airplane types were developed to accomplish this: the torpedo-bomber and the dive-bomber. The former came in at wave-top altitude and dropped a 13-foot-long, 2,000-pound torpedo within 3,000 feet of a warship. The latter approached at high altitude and then rolled into a nearly vertical dive, using specialized flap-like dive brakes to keep from building up excessive speed on the way to dropping a bomb from an altitude of 3,000 feet or less. These weapons were similar to those developed for use by ships against ships: torpedoes launched from submarines, destroyers, and torpedo boats; and high-explosive artillery shells fired from battleships, cruisers, and destroyers.

These airplanes were also assigned collateral missions. For example, the torpedo bomber was initially equipped with the Norden bombsight to aim bombs to be dropped from level flight at medium altitudes.[1] The dive-bomber was utilized as a scout to search for and locate the enemy fleet. Both bomber types were manned by a pilot and tail gunner, who was also initially the radioman when radios were added to tactical airplanes. The torpedo bomber was bigger than the dive-bomber because its standard weapon, the torpedo, was bigger and heavier and it had a three-man crew, the third crewmember being a bombardier for level-bombing missions.

Survivability was even more critical to success in carrier warfare than it was for a land-based air force. Each carrier sallied forth with a complement of pilots and airplanes that could not be readily replenished, particularly given the vast expanse of the Pacific. Even the big Essex-class carrier carried less than 100 airplanes and an increasing proportion of those were fighters after kamikaze attacks began. Successful strikes against another carrier task force were also dependent on a sufficiently large number of airplanes attacking at one time in order to overwhelm the enemy's defenses. It was the carrier equivalent of a battleship's broadside. Notwithstanding the fact that the loss of a pilot and his crew was a tragedy, the loss of an airplane was an incremental diminishment of the carrier's operating effectiveness or even its ability to survive.

Firsthand experience after Pearl Harbor quickly demonstrated the relative merit of each means of attacking a ship. The Army Air Forces had virtually no success in sinking maneuvering ships with bombs dropped from level flight at altitude, even using the vaunted Norden bombsight. The ship simply turned away during the bomb's fall and even if it didn't, the bombs almost always missed. The U.S. Navy's torpedo-bomber crews initially fared even worse, since in addition to the likelihood of a miss (or if a hit was made, the torpedo not detonating), they were very likely to get shot down. Dive-bombing, on the other hand, was effective and reasonably survivable. The ship could not easily dodge a bomb dropped from 3,000 feet or less with an imparted speed of 300 mph. Its gunners had trouble elevating their weapons high enough to hit the bombers in their dives. Defending fighters, not having dive brakes, found it difficult to make accurate gun runs on dive-bombers during the dive itself, which limited them to being high enough initially to make interceptions before the attack or taking revenge during the dive-bomber's withdrawal.

The TBF Avenger was the Navy's front-line torpedo bomber, replacing the obsolescent TBD Devastator in 1942. Here it is dropping an Mk 13 torpedo in trials to increase the speed and altitude of release. A plywood box surrounds the torpedo steering fins and counter-rotating propellers. This stabilized the torpedo in the air for proper water entry, at which point it broke away. The vapor is from the steam that was generated to turn the turbine that drove the propellers. (U.S. Navy photo via author's collection)

The U.S. Navy dive-bombers initially did not have armor-piercing bombs. The Bureau of Ordnance did not consider them to be effective because of their lesser explosive content and the need to be dropped from 10,000 feet to achieve an effective terminal velocity. A near miss, which was more likely than a direct hit even when dropped from a lower altitude, would not have the same concussive impact on the ship's hull or rudder as a general-purpose bomb of the same weight. For the Pearl Harbor attack, the Japanese had simply attached fins to 40-cm battleship projectiles, which had been machined to a more aerodynamic shape. It was a potentially devastating weapon. However, at 1,760 pounds it could only be carried by the big Type 97 Model 3 (Kate) attack bomber and when dropped in level flight at 10 to 12,000 feet, was not likely to hit a maneuvering ship. In fact, almost all missed the moored ships at Pearl Harbor. *Arizona* was an unfortunate exception. The effectiveness of the U.S. general-purpose bomb was somewhat increased by the adoption of the delay fuse, which

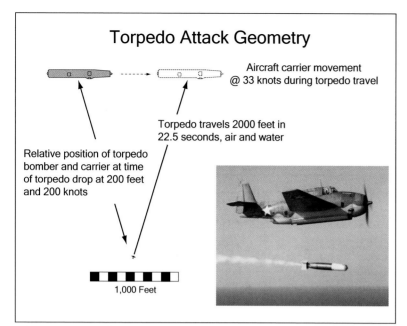

Figure 1-1. (Author)

The armor-piercing bomb—shown here waiting to be loaded in an SB2C—was not frequently used. If it hit a ship, it was more effective than a general-purpose bomb, but since it contained less explosive for the same bomb weight, it caused less damage in the near miss that was more likely. The men are adjusting the bomb displacing yoke. It attached to pins that extended out from the side of the bomb. (U.S. Navy photo via Steve Ginter)

resulted in penetration of lightly armored decks and therefore greater damage. Armor-piercing bombs that weighed 1,000 and 1,600 pounds were developed by the Bureau of Ordnance during the war but saw limited use. Semi-armor-piercing bombs were also made available. These found better acceptance because of the greater explosive charge for the weight.[2]

The Navy quickly dropped the idea of level bombing against ships but persevered with the torpedo attack in spite of early disappointments and even the tragedy at the Battle of Midway when the only contribution of the torpedo bombers was to spread the enemy's defensive firepower and fighter assets over an overwhelming number of attackers coming out of the sky from all altitudes and directions. The primary problem was the torpedo itself.

The Navy's aerial torpedo at the start of the war was the 22.5-inch diameter, 13-ft 5-inch long Bliss-Leavitt Mk 13. When fitted with a 400-pound warhead, it weighed a little less than 2,000 pounds. With a 600-pound warhead, it weighed 2,216 pounds. In the water, it had a speed of about 33 knots and a maximum range of 6,000 yards. It had been specified and developed when the Navy's torpedo bombers were biplanes, so it was supposed to be dropped at an altitude of 50 feet and airspeed of less than 125 mph. It also had to be dropped within 1,000 yards for any chance of a hit since it was unguided. Upon entering the water, it might very well take up a different heading than intended, wasting the effort required to accurately estimate the target's speed and establish the proper (and significant) amount of lead as shown in Figure 1-1.

It would have been even worse, but just before the war began, the Mk 13 was modified with a frangible open plywood box around the existing torpedo tail vanes. The box provided directional stability in the air after the torpedo was dropped that increased the likelihood of a proper water entry with no broaching. The box broke away when the torpedo entered the water.

Simultaneous attacks from both beams were almost essential against a maneuvering warship, since the torpedo was not much faster than a warship and unguided. If there was only one torpedo to avoid, the target might well maneuver out of its path or outrun it. In addition to the drop airspeed, altitude, and range limitations making the torpedo plane far too easy a target during the attack, if the early Mk 13 was dropped within limits, it was likely not to run, and if it did run, was likely to run too deep to hit, and if it did hit, would probably not explode.[3]

As a result, few torpedo attacks during 1942 and early 1943 resulted in hits and even fewer ships sunk, although the Douglas TBD pilots claimed hits on, and the sinking of, the Japanese light carrier *Shoho* in the Battle of Coral Sea in May 1942. The nadir was at the Battle of Midway the next month. None of the Navy's 47 torpedo bombers–41 obsolescent TBDs, and six brand new TBFs–that attacked the Japanese fleet hit a ship with a torpedo that exploded. Only five returned. Of the 100 aviators that took off in these airplanes, all but 15 died in the attack or its aftermath.

However, the problem was not just the poor performance of the torpedo. Most of the attacking torpedo bombers at Midway were shot down before the torpedo could be dropped. As reported in Admiral Nimitz's action report[4]:

The dive-bomber was equipped with large flaps to keep speed from building up too fast in a steep dive and a bomb displacing yoke (already retracted here) to ensure that the bomb did not hit the propeller after release in a steep dive. This SB2C is assigned to a training squadron and is dropping an obsolete bomb. (U.S. Navy photo via Steve Ginter)

- *The excellent Coordination of Dive bombing and torpedo plane attacks, so successful in the Coral Sea, was missing in the Battle of Midway. Chief among the factors preventing coordination were the Japanese tactics in concentrating fighters on our torpedo planes. This let the dive bombers in so that we sank their carriers just the same, but at the very high cost of most of our torpedo planes.*
- *TBD planes are fatally inadequate for their purpose. The loss of the brave men who unhesitatingly went to their death in them is grievous. The TBF is much improved, but still cannot attack ships defended by fighters without fighter support. Long range carrier fighters must be developed.*
- *Effectiveness of aircraft torpedoes ... must be increased. A larger torpedo warhead is urgently required. The present strengthened torpedo is a favorable step in the right direction, but the torpedo must be designed for much higher speed drops. In the Midway action the B-26 and TBF planes received their most serious losses from Japanese fighters when they slowed down to limiting torpedo dropping speed.*

Nimitz also noted, "Had the 1,000-pound armor piercing bomb under development been available for dive bombers, fewer of the many ships that were hit would have escaped; and fewer hits would have been needed to destroy the carriers."

A subsequent report on the battle reiterated the conclusion about level bombing and supported the continued employment of torpedo bombing, notwithstanding the lack of hits at Midway:

> *High-altitude horizontal bombing has proven itself relatively ineffective against maneuvering surface vessels. As Commander Cruiser Division SIX states, "Our own sea forces, and apparently enemy sea forces, have little respect for high altitude bombing, the results of which are mostly 'near misses'," and not near enough. Even in peacetime, high- altitude horizontal bombing from about 10,000 feet results in only a small percentage of hits on a maneuvering target of battleship size, and as the altitude increases the percentage goes further down. Such results will not stop a determined fleet. On the other hand, the aircraft torpedo and dive-bomber have proven themselves, in this action as well as in all prior experience of other belligerents, to be the only truly effective weapon for such attack.*[5]

It helped that the submariners were experiencing the same frustration at the lack of hits and detonations. As a result, the standard Mk 13 aerial torpedo was significantly improved during the war, receiving more than a dozen modifications. The firing mechanism and depth control defects were discovered and corrected, which enabled the submarines and torpedo planes to be devastatingly effective. A wooden shroud on the nose known as the pickle barrel and a modification of the tail box was developed and qualified for drops at 200 knots and 200 feet. The pickle barrel slowed the torpedo before it hit the water and then acted as a shock absorber. The tail box kept it at a proper entry attitude. The force of the water entry stripped the wooden appurtenances from the torpedo, which was then free to seek its pre-programmed depth and run straight.

Because of the value initially placed on the ship-sinking capability of the torpedo and to increase the strike flexibility of the air group, the SB2C and even fighters were modified to carry the torpedo. This Service Test SB2C is carrying one modified with the nose cap developed to reduce the drag of an externally mounted torpedo. (U.S. Navy photo via Steve Ginter)

The F6F Hellcat fighter was also evaluated as a torpedo bomber. Combined with a torpedo that could be dropped at high speed, this allowed for a less vulnerable delivery. This picture, dated 28 May 1943, depicts the minor modifications required to retain and stabilize the torpedo while it was being carried. The torpedo being used is a dummy. The F6F is a flight test aircraft with the early .50-cal. gun barrel fairings. (Grumman History Center)

The Navy tinkered with the Mk 13 torpedo all through the war to allow higher and faster drops. The one this ordnance team is loading on a TBM has the plywood tail box as well as the drag ring, or pickle barrel, on the nose. The drag ring slowed the torpedo if dropped at high speed and altitude so that water entry was at the lower speed that it was originally designed for. When the torpedo hit the water, the drag ring was stripped away along with the tail box. (U.S. Navy photo via Steve Ginter)

Torpedo development was also accomplished by Caltech in California beginning in 1943. The object was a torpedo that could be dropped from an airplane flying at 350 knots and 800 feet and still run true to the target. A test facility was built at a reservoir east of Pasadena where torpedoes could be dropped into the water at various speeds down a long chute. The result was a production modification with a shroud ring ahead of the propeller blades, stronger propeller blades, and a more rugged gyroscope, among other detail changes. After evaluations by VT-13 in early 1944, it was qualified to be dropped from as much as 800 feet at speeds approaching 300 knots. The first operational use was on 4 August 1944. In late 1944, the so-called ring-tail torpedoes were effectively employed by VT-51 against Japanese merchant and war ships at Okinawa and the Philippines. Radar was also used to make an initial approach in cloud or poor visibility and provide a more accurate determination of range.[6]

Concurrent with development of the torpedo, the Bureau of Ordnance worked to improve the torpedo sight. At the start of the war, the torpedo plane was equipped with the Torpedo Director Mk 28. This proved to be inadequate to the task. Even if the pilots had been provided with more experience aiming and dropping training torpedoes to see where they went relative to the ship they were attacking, it required too much time to configure and operate when torpedoes could be dropped at higher speeds. At least two improved directors were developed and fielded, but most pilots preferred to drop by eye. Use of the directors during an attack still required too much attention and imposed limits on maneuvers that pilots considered important for survivability.

By the end of the war, notwithstanding the lack of an acceptable sight, the torpedo had fully regained its place among the airplane-employed weapons. Torpedo-dropping TBM pilots could claim the lion's share of the sinking of the Japanese battleship *Yamato* in April 1945 off Kyushu. The last significant action by U.S. torpedo-carrying bombers in the war, it was also the swan song of the torpedo as an aircraft-delivered weapon for sinking surface ships. (A Japanese torpedo bomber would get a hit on—but not sink—the battleship *Pennsylvania* (BB-38) at Okinawa on 12 August.) Submarines would continue to rely on them to sink ships. Antisubmarine warfare aircraft would still be armed with homing torpedoes to sink submarines. The ability to carry them would still be required of attack airplanes for a few more years. The U.S. Navy would take aerial torpedoes out of storage during the Korean War to disable a dam, but it would never again attack a surface ship with one.

In addition to destroying the Japanese fleet, the Navy carrier-based airplanes were heavily involved in support of Marine assaults and subsequent land operations. These included close air support, infrastructure destruction, supply line interdiction, etc. Here new weapons came into play, the air-to-ground rocket and less successfully, the forerunners of today's guided missiles. Cannons also began to be substituted for machine guns because of the greater destructive power of their explosive shells. Torpedo-bombers were used in level and glide bombing against shore installations using general-purpose bombs.

These missions increasingly involved the use of fighters as bombers. Fighters had always had a collateral mission of defense suppression and close air support. They had even been provided provisions for dive bombing (Corsair main landing gear extension) and torpedo delivery (Hellcat). Air-to-ground capability became even more important when the kamikaze threat necessitated adding more and more fighters to the air group to defend the carrier. Since the number of aircraft aboard was fixed, this imperative crowded out some of the bombers, requiring the fighters to pick up the load when necessary. Rocket rails were added in addition to existing auxiliary fuel tank racks that could be used to carry and drop bombs.[7]

Radar

Originally developed to warn of approaching enemy aircraft, radar was first restricted to shore-based installation due to the size of antennas and electronic equipment. It was then refined for use on large ships and finally, small enough to be carried by patrol airplanes and carrier-based bombers. The initial presentation, though, required considerable interpretation that was hard to accomplish while also flying, particularly on instruments. It was, however, almost the equivalent of super powers. Radar meant the U.S. Navy pilots could find enemy ships in the dark while the Japanese defenders on most of those ships could not see them.

The early airborne radars could detect landmasses and ships at night and in poor weather, which made them very useful for navigation. It facilitated attack in those conditions, which increased the survivability of a torpedo drop in particular. The standard radar on carrier-based airplanes for most of the war was the AN/ASB adapted by the Naval Research Laboratory from a radio altimeter application. Its development was significantly advanced in August 1941 that BuAer issued a preliminary plan to install it in the TBF torpedo bomber and require provisions for it in new scout bombers. Twenty-five sets had been ordered from the RCA Manufacturing Company in December 1941. It had a range of about 30 miles. The antennas were Yagi arrays[8], one under each wing. These were installed in a handful of TBFs and SBDs in late 1942. One of each deployed on *Saratoga*

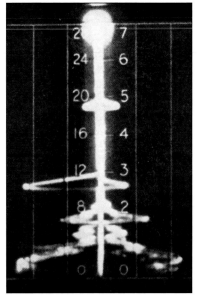

The A-scope presentation used with the Yagi arrays only indicated the distance of targets along the line that the antenna was pointing, with the left side presenting the radar return from the left antenna and the right side, the right antenna. The width of the return was indicative of the size of the target and its distance up the scale, its distance from the aircraft. Note that the antennas could be pointed outboard in two different directions but the antenna direction is not provided on the display itself. (Author's collection)

and the rest were shipped to Pearl Harbor to be deployed on other carriers. The AN/ASB became standard equipment on Navy bombers and more than 26,000 were produced during the war. It was continually improved with the last being the –8.

The two antennas were directional and were rotated by the operator to the axis of interest with the radarscope providing range to objects along that axis. The antennas could be pointed in different directions at the same time. Although the workload of antenna positioning and A-scope interpretation limited the radar's utilization by a pilot, a repeater scope was provided in the cockpit. The first night bombing attack by carrier-based airplanes using radar was accomplished in the early hours of 17 February 1944 by VT-10, flying TBM-1Cs from *Enterprise* and attacking the Japanese oiler and cargo ships in the anchorages at Truk. The crewman in the radio compartment would coach the pilot into

The TBM-3D that was deployed in 1944 featured a permanently installed APS-4 radar on the right wing and B-scope radar displays for the radioman and pilot. This provided more mission capability at night and in poor weather. Also shown are under-wing-mounted 3.5-inch rockets with 5-inch warheads added for greater offensive firepower. (National Archives 80-G-408593)

The first radar set carried by U.S. Navy carrier-based aircraft used the Yagi array for an antenna. There was one under each wing. They could be separately aimed from the radioman's position within the fuselage. This one is pointed almost directly outboard, in the direction that the pilot of the TBF is looking. (Grumman History Center)

The addition of 20mm cannons added a significant punch. This F4U-1C was produced with four 20mm cannons in place of the original six .50-cal. machine guns. (Vought Heritage Center)

position for the run-in at 250 feet to drop 500-pound bombs with four-second delay fusing into the sides of the anchored ships. A visual drop was preferred, but there were instances when it was made on a countdown from the radar operator to the pilot. Due to the point-blank range permitted by night glide-bombing attacks, the pilots claimed 50 percent of their attack runs were hits, compared to only 20 percent in daylight operations. One TBF of the 12 making the strike was lost, but in general the defensive fire was reported to be inaccurate.[9]

The development of the magnetron enabled higher frequencies to be used, up to 3,000 MHz compared to the 500 MHz of the ASB, significantly improving the effectiveness and accuracy of small antennas. The duplex antenna, which meant that the same antenna could both

transmit and receive, was another innovation that allowed compactness. One of the radars that resulted was the APS-4, which could be hung beneath the wing of any airplane on a standard bomb rack. The B-scope displays provided in both the cockpit and the rear compartment presented both range and bearing, making interpretation easier. Detection range of a ship increased to almost 100 miles. Avengers and Helldivers that could be equipped with the APS-4 were designated TBM-3Es and SB2C-3Es. These derivatives began to be deployed late in the war, and all the new bombers, including the single-seaters, were to have provisions for the APS-4.

Four carrier-based, dedicated night-attack squadrons deployed during the last months of World War II. They all used the TBM-1 or -3 D model, which was modified to have a radar pod extending from the leading edge of the right wing. Some were further modified in the field to eliminate the turret and other equipment to lighten the aircraft and permit carrying extra fuel in tanks in the bomb bay. The pilot was provided with a small radarscope, but the main radarscope was in the bombardier's compartment, where two radar operators took turns operating it. Attacks with 500-pound bombs and rockets were carried out visually after targets were found and approached using the radar.[10] (Flares were occasionally used for illumination, and searchlights had been employed in antisubmarine warfare. However, the latter provided an excellent aiming point for ground-based anti-aircraft fire.)

In addition to employing radar, the Navy modified TBMs in Night Air Group 90 with radar jammers in December 1944 to augment the release of chaff, which were strips of aluminum foil that reflected a disproportionate amount of the radar transmission and obscured the return of real targets. The jammer transmitted random noise on the threat radar's frequency. It was successfully demonstrated in January in a mock attack on *Enterprise*, making the Combat Information Center's air search radarscope completely unreadable. Chaff—although low tech, minimally effective against some radars, and effective against the threat radar only if cut to the right length for its frequency—has also continued to be used as a countermeasure to the present day. The modern-day equivalent of the foil strips is aluminum-coated glass fibers.

Armament

Early bombers were lightly armed in terms of forward firing guns. The original TBF-1 had a single cowl-mounted .30-caliber machine gun, synchronized to fire through the propeller. This was soon deleted in production in favor of a .50-cal. gun in each wing with 600 rounds per gun. The SB2C was first produced with four forward-firing .50-cal. Browning machine guns. This was soon changed in production to two 20mm cannons with 400 rpg. These SB2C-1Cs were the first American carrier-based airplanes to be armed with a 20mm cannon. The rate of fire was 750 rounds per minute and the muzzle velocity 2,800 feet per second, both about the same as the Brownings. More importantly, the bullets were two to three times heavier and could be filled with an explosive charge. Although there were some initial difficulties with feed jams, 20mm cannons became standard on new production carrier-based fighters and bombers after the war.

The most successful and lasting development during the war was the unguided rocket. Every operational carrier-based fighter and bomber was modified to carry and fire rockets beginning in late 1943. First to be armed with the rocket were TBFs flying Anti-Submarine Warfare (ASW) missions from escort carriers in the Atlantic. The U.S. version was developed from a British ASW rocket with a 3.5-inch diameter warhead and solid fuel motor. The first improvement was the substitution of the equivalent of a 5-inch antiaircraft shell for the British warhead. This reduced range and impact velocity, so the next change was the substitution of a bigger, 5-inch-diameter motor for the

A typical weapons loading scene on a carrier during World War II. The wing at the top of the picture belongs to an F6F-5N armed with six .50-cal. machine guns and six 3-inch rockets with 5-inch warheads. The TBMs are being armed with rockets and 500-lb general-purpose bombs. (National Archives 80-G-321903)

original 3.5-inch motor. The result was the High Velocity Aircraft Rocket (HVAR), 6 feet long and weighing 140 pounds. The warhead could reportedly penetrate 1.5-inch-thick armor or up to 4 feet of reinforced concrete. Pylons were provided for eight to ten rockets, meaning that each airplane carrying 5-inch rockets was in effect capable of delivering the punch of a destroyer salvo.

In order to provide even more penetration capability, a bigger rocket, *Tiny Tim*, was created by putting the business end of a 500-pound semi-armor-piercing bomb on the front of standard oil well casing (conveniently, the same 11.75-inch-diameter as the bomb) filled with rocket propellant, and adding a rocket nozzle and fixed fins to the back end. The result was 10 feet long and weighed 1,300 pounds. The first test firing took place in June 1944. It was immediately obvious that it could not be fired off a rail because of the rocket blast. The first solution was a hinged yoke like that used on dive-bombers to provide prop clearance in a steep dive. However, the lighter and simpler approach was to just drop it, with a lanyard used to initiate ignition after the missile had fallen clear of the airplane. Originally intended for use against German V-weapon bunkers, it was deployed to the Pacific instead, as the invasion of Europe was taking care of the V-weapon bunker problem. Unfortunately, the first carrier carrying Tiny Tim rockets, *Franklin*, was all but sunk by kamikazes before strikes could begin. Tiny Tim was employed with some effect against Japanese shipping by shore-based Marine Corps squadrons flying North American PBJs, the Navy's designation for the B-25.

The Tiny Tim rocket was designed to deal with heavily fortified enemy positions. Because of the rocket blast, it was dropped and the motor ignited when it reached the end of a lanyard attached to the mounting pylon. Although qualified for use during the war, this is a test firing from an F6F Hellcat in May 1948 at NAS China Lake for ongoing development. (Gary Verver collection)

Remotely Piloted Aircraft and Missiles

Primitive guided missiles were evaluated in combat toward the end of the war, with the objective of increasing both the effectiveness and survivability of the strike. Effectiveness was primarily attained by the accuracy of the delivery while survivability was achieved by keeping the launch aircraft out of the range of the target's defenses. With conventional unguided weapons, these weren't compatible. The Japanese kamikaze attacks were very accurate but not intended to be survivable: the Army Air Forces bombing runs on ships while at high altitudes were not very risky, but were almost completely ineffective.

In order to hit the target from a distance, the weapon had to be guided after its release to compensate for the target's subsequent maneuvers and/or any initial error in aim. It was best that the weapon be autonomous once the target was identified to it—launch and leave. However, this had drawbacks. Once on its own, the missile could not employ defensive maneuvering. If target acquisition were lost for even a moment, reacquisition might not occur. Autonomy also generated complexity, which decreased reliability and increased cost.

The genesis of the first generation of guided attack weapons was the development of remotely piloted targets for antiaircraft gunner training. The U.S. Navy program began in 1936. Initial remote control flights were accomplished using an NT trainer with a safety pilot aboard to take over if the system malfunctioned and for landings. Two N2C-2 biplane trainers and a pair of Stearman Hammond Y-1Ss (designated JH-1) were also converted to radio control by the Naval

Aircraft Factory.[11] In November 1937, a successful flight was made at Cape May, New Jersey, of an N2C with no safety pilot aboard. The big Great Lakes TG-2 biplane torpedo bomber was used as a control airplane with the drone controller in the front cockpit. A JH-1 was flown against the gunners of the carrier *Ranger* in August 1938 with no hits in two runs. An N2C-2 was used to dive bomb the battleship *Utah* (BB-31) in September, but was shot down on the first run. This appears to have been luck, because over-flights of several different ships in February 1939 resulted in the slow N2C drones rarely being hit and only one being shot down. Dive bombing attacks against *Utah* in March 1939 resulted in no hits in nine runs. Improvements in antiaircraft fire control were instituted as a result. The other insight was that remotely controlled aircraft would be a very effective weapon.

These drones were controlled visually from a nearby airplane, which limited the aiming accuracy of the drone as a weapon and exposed the controlling aircraft to enemy defenses. Dr. Vladimir Zworykin, RCA's lead scientist for television, had proposed an "aerial torpedo with an electric eye" to the Bureau of Ordnance in 1935. It was rejected "after a number of conferences…because the bomb would have to be of such size and weight (around 1,600 to 2,200 pounds) and the control plane of such complexity and size that the striking power of a carrier would be reduced by 50 percent."[12] In 1937, under a contract to the Soviet Union for airborne reconnaissance using television—Zworykin was a Russian émigré—RCA's camera was taken aloft in a Ford Tri-Motor and successfully transmitted usable pictures of the ground below it to a mobile unit. Although the original interest at the Naval Aircraft Factory was in televising the instrument panel of a remotely controlled aircraft doing high-risk structural tests, the result was its application to drone control, providing a picture to the drone controller of what the drone was flying toward. The feasibility began to be evaluated in flight in early 1941 with prototype equipment installed in a Lockheed XJO-3 transport. Clear images of ships were acquired at up to eight miles and transmitted 60 miles.

In the summer of 1941, a remotely controllable TG-2 torpedo bomber was equipped with a self-contained camera/transmitter hung from a bomb rack and a newly invented radar altimeter. One, with no pilot aboard, was used to make a torpedo attack on a destroyer steaming in Narragansett Bay off Quonset Point, Rhode Island. The control pilot, 20 miles away and watching a television picture transmitted by a camera mounted in the nose of the drone, guided it to a torpedo release about 300 yards from the target. This was followed later that month by a deliberate crash of a television camera-equipped remotely controlled BG-1 biplane into a towed target raft with the controller 11 miles away. These successful tests resulted in the establishment of a major assault drone program, *Project Option*, in May 1942 by the Chief of Naval Operations.

The Naval Aircraft Factory had already been authorized to design and build 100 attack drones, TDNs, powered by two Lycoming 220-hp air-cooled engines, which were not critical to the war effort. A range of 600 miles and a speed of 150 mph were required with a 2,000-pound payload, along with the capability to be catapulted from an aircraft carrier. The first TDN flew in November 1942.

The TDR/TDN control aircraft was a Grumman TBM, since the drones were originally to be carrier-based. The drone controller sat in the cockpit behind the pilot, which was normally not occupied in a TBM. At least a few TBMs were retrofitted, like this one, with an H2X radar in a retractable dome installed just aft of the bomb bay. Using a Plan Position Indicator (PPI) display and with a transponder in the TDN/TDR, the controlling pilot in the TBM could fly the drone to the vicinity of the target without being able to see either one. All that was necessary to complete the attack was for the drone to be aimed at the target while close enough to see it on the TV monitor. (National Archives 80-G-387191)

The TDR mockup was reviewed on 23-24 April 1942. It may look a little crude, but the configuration didn't even exist as a three-view drawing before 18 March, when Interstate Engineering's management was asked if it was interested in building an assault drone. Also displayed are some of the weapons that it could carry, including depth charges and the Mk 13 Torpedo. The cockpit was for maintenance and ferry flights only. The opening in the nose is for the TV camera that provided guidance to the controlling aircraft. Although lightly built and only powered by two 220-hp engines, the aircraft was not small. It was more than 36 feet long with a 48-foot wingspan. (National Archives)

In April 1942, BuAer provided a letter of intent to the Interstate Aircraft and Engineering Company, which was located in Los Angeles, California, for the TDR. It was very similar to the TDN. The major difference was that the wing was mounted low on the fuselage instead of high as on the TDN. The gross weight of the TDR with an Mk 13 torpedo and 100 gallons of fuel was about 6,700 pounds. It was projected to have a maximum range of 750 miles, reduced to about 400 miles when flying at its top speed of 160 mph. Interstate subcontracted much of the airplane to manufacturers with no aviation experience. In part as a result, only 188 were produced of the 1,000 TDR-1s, TDR-2s, TD2Rs, and TD3Rs eventually ordered.

A rudimentary cockpit was included in the drones for initial flight test, drone control development, and ferry when required. On its last flight, the landing gear was dropped after takeoff to maximize speed and range. (A 100-pound bomb was also to be carried in the cockpit, probably to insure destruction of the control equipment.) The drone had a simple autopilot and radio altimeter for altitude hold with the controller having a control stick to change its course and altitude and a function control that was simply the dialer from a rotary telephone. In combat, the drone operator was located in an accompanying Grumman TBM, using video from a TV camera and radio controls to remotely control the TDN or TDR. The TBM had a four-man crew, adding a drone pilot to the standard complement of pilot, tail gunner, and radioman. The pilot in the cockpit flew both the TBM

Satisfactory demonstrations were accomplished in early 1943 at Cape May, New Jersey, with the TDN. Carrier takeoffs with no pilot aboard were accomplished from the training carrier Sable (IX-81) operating in Lake Michigan in August 1943. For some reason this particular NOLO (No Local Operator) takeoff was toward the stern with the ship backing down. One explanation for this is that it provided the longer deck run with no risk of the drone veering off into the island in the event of an engine or control failure. (National Archives 80-G-387174)

Higher-performance drones were ordered from Interstate. The TD2R was to be powered by two 450-hp Franklin O-805-2 engines, but it was canceled before the prototypes were completed. The TD3R was powered by two 450-hp Wright R-975 engines to allow an increase in gross weight to 10,343 pounds, which allowed an increase in fuel for a range of 1,250 miles. This XTD3R is armed with an Mk 13 torpedo. The cockpit was again only for acceptance, maintenance test, and ferry flights. Four TD3R prototypes were built before production was canceled by mid-1944. However, at least one was flown and evaluated in March to June 1945 trials that culminated in a strike on an abandoned lighthouse on Waugoschance Point, Michigan, on 8 June 1945. (Author's collection)

and the drone initially. The drone pilot in the cockpit behind the pilot took over when the drone was near the target and it was visible on his TV screen.

As an alternative to remote control by reference to a television picture, a homing device was created from a radar altimeter aimed horizontally. This was also potentially an all-weather targeting system. It also allowed for automatic release of the torpedo when in range. Evaluated in flight on a drone in mid-1943, it was able to acquire a moving ship from a distance of two miles and fly directly over it. However, television allowed an attack on less well differentiated targets and remained the operational targeting system.

The assault drones were to be operated by Special Task Air Groups (STAGs). Three were formed at NAS Clinton, Oklahoma. After initial training there, STAG 1 moved to Traverse City, Michigan, to conduct overwater training. It then repositioned to Naval Auxiliary Air Station Monterey, California, to be ready for deployment. STAG 2 relocated to NAS Eagle Mountain Lake, Texas, after training at NAS Clinton. STAGs 2 and 3 were decommissioned in March 1944 with STAG 2 being integrated organizationally with STAG 1 and some of STAG 3's personnel and aircraft transferred to the Special Weapons Test and Tactical Evaluation Unit (SWTTEU) that had been established at Traverse City, Michigan, in August 1942.

There was a significant lack of enthusiasm in the fleet for substitution of drones for manned strikes launched from an aircraft carrier. For

one thing, the TDN and TDR were thought to be too slow for combat effectiveness.[13] For another, a drone took up significant space (folding wings had been thought to be an unnecessary complication and expense) and could only be used one time, whereas conventional manned airplanes could be used to deliver bombs many times, providing a greater strike capability for the fixed volume of the aircraft carrier. Replenishment with the large and fragile drones while underway was much more onerous than replenishment with bombs. These shortcomings were not enough to overcome the argument that a single drone was more likely to make a hit with a bomb or torpedo than several manned airplanes, since the attack would be pressed to point-blank range.

As a field trial operating from shore bases, not carriers, two STAG 1 squadrons deployed to the Russell Islands in the South Pacific in June 1944 with TDR-1s and TBMs. A practice mission was flown on 30 July against an abandoned Japanese freighter grounded near Cape Esperance on Guadalcanal. Four TDRs, each armed with a 2,000-pound bomb, were successfully launched. Two were direct hits, and the other two were near misses.

In September, operations commenced in earnest against Japanese targets on and near Bougainville and Rabaul. Over the space of 30 days, 46 TDRs were launched, with 37 getting at least as far as the target area, and at least 21 making direct hits. The Japanese initially believed that the Americans had begun to make suicide attacks. Although arguably successful as employed, the concept continued to

The Bat was a compact guided missile that was unpowered. A radar in the nose allowed it to home in on the target autonomously. It carried a 1,000-lb bomb suspended from the wing spar in its "belly." The only control surfaces were on the wing. The empennage was for stability only. (Terry Panopalis collection)

The Bat and the Pelican were light enough to be carried by carrier-based aircraft but far too large to fit in even the Avenger's bomb bay. In this case the Bat is being carried externally on a TBM with its vertical fins folded for ground clearance. (National Archives 80-G-703162)

be regarded as operationally ineffective by senior Naval officers. The major problem was the crude resolution and poor contrast provided by the TV camera. It was only useable on a bright day against a distinct target. Additionally, the fragility and low speed of the TDR made it susceptible to being shot down during the attack.

Even if STAG 1's 30-day trial had been an unqualified success, its future was predetermined. OpNav had concluded that guided missiles "should be designed for use by carrier-based aircraft with the minimum of alteration or modification of these planes."[14] STAG 1 was withdrawn from combat operations at the end of October 1944 and decommissioned in December. The remaining TDRs were primarily employed as targets at the Naval Ordnance Test Station at China Lake, California.[15]

In recognition that accurate guidance with the crude television picture was problematical except in daylight and good weather against a target that contrasted with its background, programs had been initiated to provide terminal guidance by radar. Two were fully developed, the Pelican and the very similar Bat, both gliding bombs that had a range of four or five miles for every 5,000 feet of release altitude. The primary difference was the guidance system. The Pelican, like later air-to-air missiles, incorporated a semi-active radar seeker. It homed on the radar reflections from the target that was being illuminated by a radar carried by the control airplane, in this case a PV-1 land-based patrol bomber. The Bat carried its own radar and could proceed autonomously toward a target.

The Pelican weighed about 1,000 pounds, including a 500-pound bomb. Pelican development began in June 1942 and it was ready for drop tests in December 1942. An order for 3,000 was placed in October 1943. A rocket-propelled version was also being planned. Initial service trials in June 1944 did not go well, with eight drops

resulting in eight misses. Part of the problem was the difficulty in keeping the target continuously illuminated by the airborne radar. Although hits were obtained in subsequent tests, the Pelican was terminated in September for all practical purposes.

The Bat had an airframe similar to the Pelican's. It weighed 1,700 pounds including a 1,000-pound bomb carried internally. Testing began in May 1944. Although small enough to be carried by carrier-based bombers like the TBM and SB2C, it was to be operationally deployed with the big, four-engine, land-based PB4Y patrol bomber. It was most effective against an isolated target such as a ship, since the radar was easily confused by the presence of other objects in its field of view.[16]

Some Pelicans were converted to Moths, which were the first of the anti-radar missiles. The Moth (as in attracted to the flame) was equipped with a direction finder that could be tuned to enemy radar frequencies and coupled to the airframe's guidance system. A 650-lb warhead was included. However, none were apparently used operationally. (Scott Pedersen collection)

The F4U Corsair proved to be an excellent fighter-bomber. It was fitted first with a field-designed center pylon and then with a stores pylon on each inboard wing, as shown here on an F4U-1D carrying two 1,000-lb bombs. (Vought Heritage Center)

After reasonably successful testing in late 1944, the Bat was deployed to the South Pacific in early 1945. Operational missions began in April 1945 with limited success due to missile unreliability, inadequate training, and few suitable targets. About 3,000 Bats were produced, but most were not expended. It remained in the Navy's inventory after the war and was modified to eliminate the "hunting" caused by the control system. The improved accuracy was successfully demonstrated in early 1948 in a series of test drops against a barge equipped with a corner reflector to provide a good radar signal. However, in July 1948, four were dropped on the battleship *Nevada*, which was to be sunk off Hawaii in a fleet exercise. The first one spiraled in immediately after it was dropped; the other three missed by margins of 600 to 1,000 yards. Apparently the more complex radar signature of a real ship and the presence of other radars in the operating area confused the Bat's primitive radar.

The End of the Beginning

By the end of the war, the U.S. Navy had eliminated its principal *raison d'être*, the enemy fleet. The next biggest navy was a very distant second in size and an ally to boot. The Army Air Forces had dropped the atomic bombs that were perceived by many to have not only ended the war, but to be the only weapon that the nation needed to win the next one. Since there was little likelihood of another threat to sea lines of communication other than submarines, a large power-projection Navy was also no longer considered necessary by then. Even Jimmy Doolittle, by then a Lieutenant General, who had relied on an aircraft carrier to get him close enough to Tokyo in 1942 to accomplish his famous bombing mission, testified to the Senate

Committee on Military Affairs in 1945: "I feel (that the aircraft carrier) has reached its highest usefulness now and that it is going into obsolescence… As soon as airplanes are developed with sufficient range so they can go any place that we want them to go, or when we have bases that will permit us to go any place we want to go, there will be no further use for aircraft carriers."

Two things happened to bring the U.S. Navy back from the brink of irrelevence on the world stage and being diminished to a coast guard role. The first was proactive—it quickly developed and publicized its capability to deliver the existing atomic bombs from aircraft carriers. The second was only a matter of time—a geopolitical crisis requiring a non-nuclear military response by the United States. There was no lack of opportunity for one. The first happened to be the invasion of South Korea by North Korea in June 1950. The U.S. president, Harry Truman, elected to help counter the invasion and the Navy's carriers proved to be essential to the U.S. military effort, particularly in the first few weeks of the war.

All of the carrier-based aircraft in combat at the end of the war were the result of development programs begun before December 1941. The Navy's principal torpedo bomber was the TBM Avenger, receiving its popular name in early 1942 as a reminder of Pearl Harbor. The corresponding dive-bomber was the SB2C Helldiver.[17] Both of the Navy's carrier-based fighters, the F6F Hellcat and F4U Corsair, were also utilized for close-air support as fighter-bombers. During the war, the Navy had initiated several new fighters and bombers as well as weapons. For various reasons, none of the resulting airplanes had reached the fleet by the end of the war except for the F8F Bearcat, and it did not see combat. However, one of the Navy's World War II bomber programs was to result in a cornerstone of U.S. Navy attack aviation for many years.

CHAPTER TWO

THE NAVY'S
FIRST ATTACK AIRCRAFT

A picture of a strike aircraft surrounded by a display of armament is an iconic depiction of its capability. The caption usually did not mention that it could only carry a few of the items at one time. This example taken of an early production AD-4 in November 1949 is therefore noteworthy because it is not misleading. Even carrying its full internal fuel of 2,280 pounds (seven 55-gallon drums), a 2,200-lb torpedo, two 2,000-lb bombs, 12 5-inch rockets (1,680 pounds total), two 20mm guns, and 240 pounds of ammunition, the Skyraider was still under its maximum gross weight of 25,000 pounds. (Ed Barthelmes collection)

During World War II, the Navy initiated an alphabet soup of carrier-based bombers. None of them reached the fleet during the war. In fact, only two were produced in any significant quantity, with only one being used operationally for more than a few years. One reason so many were started and few were successful is that the Navy's requirements evolved during the war. The one that did serve, the ultimate product of all that effort, was a workhorse that deployed with air groups and air wings for more than 20 years.[1] Even after being replaced in the Navy, it was used in shore-based close-air support and combat rescue escort missions because of its payload and endurance.

The Navy's designation system was both mission and manufacturer based. An example is F8B. The first letter stood for the basic mission, in

this case Fighter. The second letter identified the manufacturer, Boeing. The number in between the two letters was the quantity of different designs of the basic mission type that the Navy had procured from this manufacturer. In other words, the F8B was the eighth different fighter that Boeing had designed for the Navy. (The first was the FB since the number 1 was not used for the initial type from a manufacturer.)

The letter T stood for torpedo, B for bomber, and S for scout. An SB was therefore a scout-bomber. The SB was traditionally capable of delivering a bomb in an extremely steep dive. A TB was a torpedo-bomber while a BT was a bomber-torpedo, the difference being that a TB's primary mission was the delivery of a torpedo, whereas the BT's primary mission was dive bombing.[2] While a TB could also drop

The SB2D was a very complicated and heavy airplane, intended as a replacement for the SB2C. Two 20mm cannons replaced the more usual .50-cal. machine guns. Only two were built before its shortcomings and revised thinking about the Scout Bomber mission and configuration resulted in termination of the program. (U.S. Navy via author's collection)

Many of the Navy's World War II carrier-based bomber programs featured the big, new Pratt & Whitney R-4360 engine. None combined with a counter-rotating propeller to eliminate torque, including the BTC-2 shown here, were successful. (U.S. Navy via author's collection)

bombs, it was limited in dive angle because it was not fitted with dive brakes like the SBs and BTs.

The manufacturers for the preponderance of Navy bombers and their designations were:

- Boeing   B
- Curtiss   C
- Douglas   D
- Grumman   F
- Kaiser   K
- Martin   M

World War II Projects

Douglas received a contract for a new scout-bomber in June 1941 to be powered by the new 2,300-hp Wright R-3350. Designated the SB2D, it was full of innovative features, from the tricycle landing gear to the upper and lower remotely controlled turrets, each housing a .50-cal. machine gun. It had a bomb bay and laminar flow airfoils to increase top speed and maximize range. Unfortunately, it wound up 2,500 pounds overweight with Douglas and the Navy sharing equal responsibility. Only two were built, the first finally flying in April 1943. The SB2D program was terminated in June 1944 and the two airplanes transferred to NACA's Ames Aeronautical Laboratory in California.

Before the war, it has been assumed that the bombers might have to attack enemy ships unescorted, with the fighters' role being to defend their own carriers. The bombers were therefore equipped with both forward and rearward firing guns, the latter being aimed by an enlisted gunner. In late 1941, only a few months after Douglas began work on the SB2D, it was becoming apparent that the addition of armor, self-sealing fuel cells, more defensive armament, and radar would overload existing dive-bombers. It was also likely that they would have to be escorted rather than rely solely on their defensive armament. The Scout Bomber class desk at BuAer, LCDR J. N. Murphy, therefore recommended that the defensive armament be removed from the dive-bomber requirement and the scouting mission—by its nature unescorted—be accomplished by the torpedo bombers, which would continue to have tail guns. For maximum mission assignment flexibility, new dive-bombers would have the capability to be armed with an externally carried torpedo and assigned the designation BT.

In February 1942, BuAer implemented a replacement program developed by the Scout Bomber and Torpedo Bomber class desks. Curtiss received a letter of intent in June 1942 for two versions of a single-seat, high-performance dive-bomber: the BTC-1 powered by the R-3350 and the BTC-2, with the new 3,000-hp Pratt & Whitney R-4360 and a different wing. Douglas was asked to develop a twin-engine horizontal bomber, scout, and torpedo plane. However, Douglas opted for a single-engine configuration powered by the R-4360, which became the TB2D, so Grumman was given the task of doing studies on the twin-engine configuration instead.

At Curtiss, priority on the SC-1 Seahawk and problems with the SB2C-1 resulted in the BTC-1 falling behind schedule and being canceled in late 1943 before any were built. The BTC-2 didn't fare much better, with first flight not until January 1945 due to basic design problems with the tail. Even then, it had to be flown with a BTC-1 wing because the bigger BTC-2 wing with a "duplex" flap system had to be redesigned following wind tunnel test. Vibration problems with the counter-rotating propeller delayed flight test and with simpler, more up to date designs in development, only two XBTC-2s and a limited flight evaluation resulted.

In place of the -1, Curtiss proposed an R-3350-powered variant of its SB2C-5 in 1944. The bomb bay was revised to completely enclose an Mk 13 torpedo. Since it used the basic structure of the SB2C, development of the resulting BT2C was expected to be quick and trouble-free. A crew position for a radar operator was retained in the fuselage because of the demonstrated importance of radar in finding and striking targets, particularly at night and in poor weather, and the high work load to use the radar effectively. Ten BT2Cs were ordered in February 1945, and the first one flew in August 1945. While an improvement over the SB2C, it was inferior to at least two of its competitors and production was terminated after nine were built.

The TB2D disappointed as well. Although Douglas received a letter of intent in November 1942, the huge XTB2D did not fly until May 1945. Also powered by an R-4360 with a counter-rotating propeller, it had a wingspan of 70 feet and maximum gross weight of almost 35,000 pounds when carrying four torpedoes. It was to be manned by a crew of three. Upper and lower turrets with two .50-cal. and one .50-cal. machine guns, respectively, provided defensive firepower. The originally proposed bomb bay was deleted during detail design in favor of four external stores stations with a capacity of 2,000 pounds each. Unfortunately for the TB2D, ongoing combat experience in the Pacific had led the Navy to conclude that single-seat bombers were preferable to multi-place bombers, the deletion of the crew and defensive armament translating into more range and payload in a smaller airplane.

In the meantime, Boeing somehow got a contract in April 1943 with few specifications and minimal BuAer oversight to develop a big single-seat, long-range fighter powered by the R-4360. Although the F8B had a fighter designation, it had a bomb bay and according to Boeing, was a "five-in-one" airplane, capable of being a fighter, interceptor, dive-bomber, torpedo bomber, or horizontal bomber. Maximum gross weight was 23,900 pounds. Three were built and flown, the first in November 1944. With the installation of a drop-

The TB2D was the huge torpedo-bomber sibling of Douglas' overly complex SB2D, with the added handicap of the R-4360 and a counter-rotating propeller. It shared the SB2D's fate. (U.S. Navy via author's collection)

pable fuel tank in the bomb bay, it could be ferried 2,800 miles or have a combat radius of 890 miles. However, the advent of the jet engine made it obsolete as a fighter. After evaluations by both the Navy and the Army in 1946, neither considered it better than other bomber options available, so it was not put into production. Trying to be jack-of-all-trades resulted in it being of interest to none.

Grumman submitted the requested study for a twin-engine torpedo bomber, its Design 55, in late December 1942, and a proposal in March 1943. They received a letter of intent on 6 August 1943 for two XTB2F prototypes and the usual engineering and test data. The engines were to be P&W R-2800-22s. The mockup review was

Close comparison of this BT2C with the SB2C will reveal that it utilizes much of the same airframe in addition to being powered by the same engine. The rear gunner position has been replaced by a radar operator's hideaway in the aft fuselage. (U.S. Navy via author's collection)

The Boeing F8B was yet another attempt to take advantage of the power of the R-4360 and avoid torque and p-factor problems with a counter-rotating propeller. This early version of a multi-role strike fighter had a small bomb bay and could carry a torpedo externally. (U.S. Navy via author's collection.)

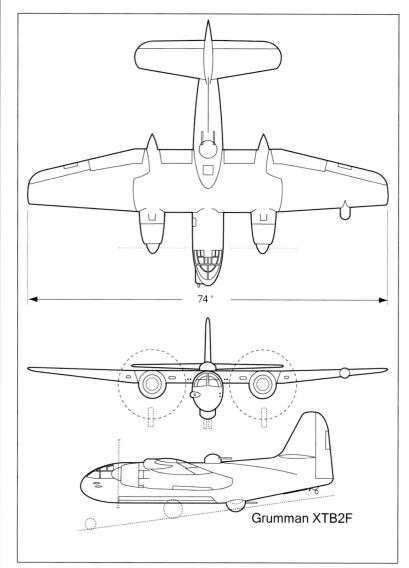

74'

Grumman XTB2F

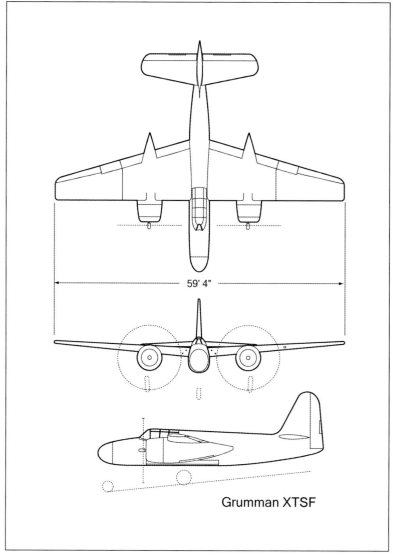

59' 4"

Grumman XTSF

The Grumman-proposed follow-on to the TBF Avenger, the TB2F, was a big, armored, and heavily armed torpedo bomber designed to allow its crew of three to fight its way, unescorted, to a torpedo drop at a capital ship. It faired even worse than its counterpart at Douglas, the TB2D, by not even making it off the drawing board. (Author)

The Grumman torpedo bomber design based on the F7F evolved to add radar and a radar operator for all-weather operations. The resulting XTSF also failed to make it off the drawing board like the TB2F. The F7F qualification was not going well, and the Navy was more receptive to a single-engine solution, possibly with jet-engine augmentation. (Author)

accomplished in May 1944. Like the XTB2D, it could carry an 8,000-pound load of bombs but at an even heavier gross weight, up to 45,000 pounds, and with an even bigger wingspan, 74 feet. BuAer decided that the aircraft was too large and heavy for operation from even the new Midway-class carriers and issued a stop work order in June. However, it authorized Grumman to continue the twin-engine torpedo bomber effort with an F7F-based concept, Design 66. It had a second cockpit like the night-fighting F7Fs, a bomb bay big enough for a torpedo, and an SCR-720 radar in the nose. In August, the XTB2F contract was revised to cover two XTSFs. Smaller than the

XTB2F but powered by the same engines, it was to have a gross weight with one torpedo and 700 gallons of fuel of 26,000 pounds. According to the specification, it was to be capable of "dives up to 50 degrees below horizontal for torpedo attack." The mockup inspection was accomplished in October, and detail design continued through the end of the year. BuAer canceled the program in January 1945 because, according to one report, Grumman was being stretched too thin. The difficulties that the Navy and Grumman were having in qualifying the F7F for carrier operations at the time might have been a consideration as well.

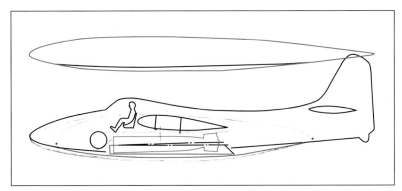

As the war progressed, torpedo bomber requirements evolved. Now the emphasis was on speed rather than defensive firepower. This is a Grumman predesign sketch of the fuselage changes required to allow the Grumman twin-engine F7F to carry a torpedo internally. (Grumman History Center)

As the war in the Pacific progressed, the mix of fighters and bombers evolved to better match defensive and offensive requirements. In March 1943, Essex-class carrier air groups typically consisted of 21 fighters, 36 scout bombers, and 18 torpedo bombers. In October, the number of fighters was increased to 36. A year later, the number of fighters in the air group was increased again, to 72, and the complement of scout and torpedo bombers reduced to 15 of each

type. However, the fighters had now been fitted to drop bombs, shoot rockets, and in the case of the F6F, even carry-and-drop torpedoes, so the strike capability had not been significantly diminished.

Douglas had redesigned the SB2D for production as the BTD beginning in April 1942, removing the second crewman and the turrets in accordance with the evolving mission requirement. The nose gear, bomb bay, and laminar flow airfoils were retained. The first one flew in March 1944. However, Ed Heinemann of Douglas was concerned that it was clearly not superior to competing airplanes. In a bold move at a program review meeting in Washington in June 1944 with production already underway, he asked BuAer to consider the development of an all-new design, preferably powered by the P&W R-2800. BuAer replied, "Maybe," but instead of the 30 days Douglas wanted for a design study, requested one that retained the Wright R-3350 to be available for review the next morning.

Literally overnight, in a Washington hotel room, Ed and two of his associates laid out the configuration and estimated the performance of what was to be the BT2D.[3] In order to minimize weight, the nose gear and bomb bay were eliminated and the wing structure and fuel system were simplified. The bomb-displacing trapeze (needed to keep the bomb from hitting the propeller when released in a steep dive) under the fuselage was replaced by a lighter approach—an explosive charge that pushed the bomb away. To provide low-speed lift, more conventional airfoils were substituted. On 6 July, BuAer authorized Douglas to end BTD production in favor of the design and

The Douglas BTD was a simplification of the SB2D design to eliminate the second crewman and the remotely controlled defensive turrets. It was a step in the right direction, as far as the evolving Navy requirements were concerned, but yet another step would be necessary. Only 26 were produced, none of which reached operational squadrons. (Robert L. Lawson collection)

The XBTD-2 was an attempt to increase its survivability by adding a jet engine to increase its top speed by 50 mph and thereby compensate for the elimination of the tail gunner. The Westinghouse 19A jet engine was mounted in the aft fuselage at an angle with its inlet located just aft of the canopy. First flight was accomplished in May 1944. The hoped-for speed increase was apparently not realized. Payload and range also suffered. In any event, Douglas was about to segue from the BTD to the BT2D. A second conversion was canceled.

The Kaiser BTK was intended to operate off the smallest aircraft carriers. This flight-test aircraft is carrying an APS-4 radar pod under its right wing, a 1,000-lb bomb on the centerline, and a fuel tank under its left wing. Note the excellent pilot visibility forward and down, a hallmark of carrier-based aircraft. (National Archives 80-G-369206)

The Douglas XBT2D-1 was the final step in the creation of the iconic AD Skyraider as the successor to both the SB2C and TBM. There are refinements still to come, but the basic configuration has now been defined and is in flight test—single-seat, Wright R-3350 engine, all-external stores, and three huge dive brakes. It was originally named Dauntless II. (Jay Miller collection)

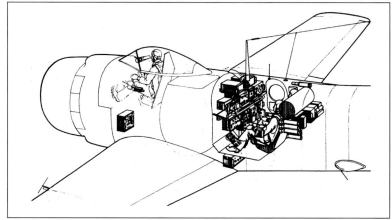

The workload required to operate and interpret the early radar during a night-attack mission was too much for a pilot. This XBT2D-1N illustration depicts the position for two radar operators in the fuselage. The pilot was provided with a repeater scope. (U.S. Navy via author's collection)

manufacture of 15 XBT2D airplanes. In an extraordinary effort, the first flight of the all-new design was made in March 1945, only nine months after BuAer agreed to the redirection. A total of 25 XBT2Ds were built, including two as a prototype of a three-seat night-attack variant, one -1P photoreconnaissance prototype, and one -1Q electronic countermeasures (ECM) prototype. Service test was accomplished in early 1946 at NAS Alameda.

Kaiser was successful in proposing a relatively small single-seat dive-bomber for a 1943 requirement to be powered by the Pratt & Whitney R-2800 engine and operated from the escort carriers. Like the SBD it was intended to replace, it did not have a bomb bay. It was initially designated BK when two prototypes were ordered in January 1944. Twenty production airplanes were ordered in October 1944. With the addition of a torpedo-delivery requirement in February 1945, it was redesignated BTK. Its first flight was in April 1945. Five airplanes were produced and flown before the contract was terminated in September 1946 in favor of other, more capable airplanes.

Martin also received a contract for a BT type in September 1943. The Pratt & Whitney R-4360 powered it. Unlike most of the other bombers powered by this big engine, it did not have a counter-rotating propeller, which may have contributed to its somewhat greater success.

It also did not have a bomb bay, which was becoming a BT trend. The BTM first flew in August 1944.

As if the programs in progress were not enough, Grumman received a letter of intent in October 1944 and a contract in February 1945 for a torpedo bomber type. In addition to carriage of the torpedo internally, higher speeds in a low-level attack were to be attained by adding a jet engine's thrust to that of the reciprocating engine's propeller. Designated the TB3F, the -1 was to be powered by an R-2800-34 engine and a Westinghouse 19XB (J30) jet engine while the -2 was to have even more power, a Wright R-3350-26 and a Westinghouse 24C (J34). There was a crew of two but no defensive armament. The second crewman was to be a radar operator and bombardier sitting beside but slightly aft of the pilot under a single canopy. The first flight of the -1 configuration was made in December 1946 but by then the Navy had reconsidered the usefulness of a dedicated torpedo bomber.

Guided Weapon Development

In addition to several bomber programs, the Navy funded guided-weapon programs. Unlike the bombers, two reached operational status during the war, as described in Chapter 1. The following did not but were part of a development process that eventually resulted in operational standoff weapons.

Gorgon was initiated as a television-guided, jet-powered, air-to-air and air-to-surface missile in July 1943 at the Naval Aircraft Factory. It became a family of winged-missiles—powered by turbojet, ram/pulsejet or rocket engines—for various missions. Most were built and tested, but none were used operationally, except as target drones. The Gorgon II series were canards and III, conventional aerodynamically. The IIB and IIIB were for the air-to-surface mission, with the IIB having Pelican-type radar and the IIIB, television guidance. Both were to be powered by a 9.5-inch-diameter Westinghouse turbojet engine. Delays in its development resulted in the elimination of these two variations from the missile program although the IIIB was used as a target drone, the TD2N/KDN.

The final attempt at making the Gorgon into an air-to-surface weapon resulted in the unpowered 2,600-pound Gorgon V. It was intended to carry the Aero 14B all-purpose tank which was a chemical weapons dispenser. Equipped with an autopilot and radio altimeter, it was to be dropped from 35,000 feet and dive to an altitude of a few hundred feet above the ground. The near-sonic speed attained in the dive would allow it to spray approximately 100 square kms before it stalled and crashed. After a few test flights to evaluate flight stability and the guidance and control system, the program was canceled.

Gargoyle was a rocket-boosted winged missile. McDonnell received a contract for five prototypes and 395 production LBDs (later changed to KSD when it became a target drone) in September 1944. Again, it was to weigh less than 2,000 pounds, including a 1,000-pound bomb. In this case, it was to be dropped from a carrier-based airplane and powered by a jet-assisted takeoff (JATO) rocket. Control was very similar to the later Bullpup missile, in that it would be

The Martin XBTM/AM Mauler was the only Navy carrier-based bomber to beat the R-4360 jinx, possibly because it didn't have the complex counter-rotating propeller. Nevertheless, its operational career was short-lived. (National Archives 80-G-70174)

The Grumman XTB3F represented yet another powerplant arrangement, in this case combining a reciprocating engine in the nose with a jet engine in the tail. The inlets in the wing leading edge are for the jet engine. In this case the radar operator was to sit beside the pilot, foreshadowing the seating arrangement in the Grumman A-6 Intruder. (Grumman History Center)

This TD2N was originally designated the Gorgon IIIB, the III signifying that it was a conventional (non-canard) airframe and the B that it was powered by a tiny Westinghouse turbojet engine. First flight was in August 1945. Only nine were delivered before it was canceled in March 1946. (National Archives 80-G-189126)

The Taylorcraft LBT-1 is one example of the straight-forward conversion of a light, single-engine, general aviation airplane to be a bomb-carrying, TV-guided, unmanned glider. The nose has been modified to eliminate the engine and add a tow hook for takeoff and cruise along with a camera for final guidance. A drag brake was added to the aft fuselage for control of dive speed and angle. The tricycle landing gear was substituted for the less directionally stable tailwheel configuration. (U.S. Navy via author's collection)

One unusual aspect of the Glomb concept was that it was to be catapult launched along with the tow aircraft when operated from aircraft carriers. Shore-based catapult trials with an F4F Wildcat (as shown here with a LNT-1) were successfully accomplished. The drawback, in addition to providing storage for a big, one-shot weapon, was the open deck space required behind the tow plane to the glider because the tow cable needed to be fully extended to minimize the snatch load on the cable and its attachments to the tow plane and glider. (William Norton collection)

remotely piloted, with the Gargoyle having a flare in the tail to permit tracking. Testing began in March 1945 with an unsuccessful drop from an SB2C. The first successful flight occurred in May 1945. Development continued into 1947 and 200 airframes were delivered, but it was never qualified as operational.

Glomb was the name created for glider bombs, a program initiated in April 1941 at the Naval Aircraft Factory. These were to be towed behind airplanes to the target area and then released and remotely con-

The McDonnell Gargoyle was a rocket-boosted, V-tailed missile developed as an air-to-ground weapon to be launched from carrier-based airplanes. It weighed 1,650 pounds and was a little less than 10 feet long. (U.S. Navy via author's collection)

trolled for the attack. Guidance was to be provided by a television picture transmitted from a camera in the Glomb's nose, although testing was also accomplished with Pelican-type radar and simple targeting radar developed from a radar altimeter. Developing an autopilot that would maintain formation behind the tow plane was a major challenge since it was impractical to do this by remote control. (This is the hardest task a glider pilot has to learn.) It involved substituting input from the towline for the visual picture that the human tries to maintain through appropriate manipulation of the controls.

Although existing Navy LNS-1 sailplanes were used initially to evaluate the feasibility of the concept, its wing loading was too light for a good simulation of the Glomb flight characteristics. The next test vehicle was a BG biplane dive-bomber with the engine replaced by remote-control equipment and ballast. This unwieldy ersatz glider was replaced in October 1942 with an Army TG-6, a glider conversion of a Taylorcraft light plane that more closely resembled the planned Glomb configuration. The Navy designated it the XLNT-1. It was modified to have a tricycle landing gear for ease of takeoff and landing. A successful takeoff and "autotow" without any assistance from the safety pilot was finally accomplished in April 1943. In a September 1943 trial flown out of Patuxent River, two XLNTs were expended against targets. In each case, the tow was successful. However, one attack failed after release due to an unusable TV picture and the other, a power supply malfunction.

The Taylorcraft XLNT/LBT was created by making only the absolutely necessary changes required to convert a propeller-driven light airplane into a remotely controlled glider bomb.[4] It had a maximum

gross weight of 5,000 pounds when loaded with a 2,000-pound bomb. The Army transferred a total of 34 TG-6s to the Navy, which were converted into LNTs. Taylorcraft also received a contract to build 25 LBTs in late 1943. The first one was flown in March 1944 and delivered in April. These LNT/LBTs were still being used in missile-guidance testing as late as July 1945 at Traverse City, Michigan.

The Naval Aircraft Factory designed a 7,000-pound Glomb with a payload of a 4,000-pound bomb; production of the NAF design was awarded in September 1943 to Pratt Read, designer and builder of the Navy's LNE-1 training glider, as the LBE-1. Piper also received contracts to design and build 100 Glombs, designated the LBP-1. It resembled a high-wing version of the NAF/Pratt Read LBE-1. At least one was completed and flown in April 1945 before the contract was terminated in June, after having been reduced to 35 in February due to lack of progress. The LBP had also been flown in April and delivered to Patuxent for evaluation in May. The LBE program was terminated in August, with only four of the 35 remaining on contract being completed.

Although each of the various elements of the Glomb concept was successfully demonstrated, and one test of an XLNT-1 at Lakehurst, New Jersey, in April 1944 using an "expeditionary team" resulted in a direct hit on the target with the tow/control airplane five miles away, Glombs weren't used operationally.

In August 1946, the CNO directed that the term guided missile be used to refer to what the Bureau of Aeronautics called pilotless aircraft and the Bureau of Ordnance called Special Weapons Ordnance Device (SWOD). In April 1947, the Army and the Navy agreed on a standard nomenclature for guided missiles. The letters A, S, and U for air, surface, and underwater, respectively, would be combined with M for missile to indicate its role. A surface-to-air missile would be a SAM and receive a name from mythology. Air-to-surface missiles, ASMs, would be named for birds of prey. Air-to-air missiles, AAMs, would be named for other winged creatures, such as Sparrow. Surface-to-surface missiles, SSMs, would be taken from astronomy.

Bomb Sight Development

Dive bombing with a near vertical approach and a low-altitude release of the bomb made accuracy relatively easy to attain. However, cloud cover sometimes precluded the necessary approach at altitude, and closeness was a two-way street from the standpoint of the target's defensive fire. Moreover, dive bombing required the employment of dive brakes to prevent the buildup of speed in the dive, and

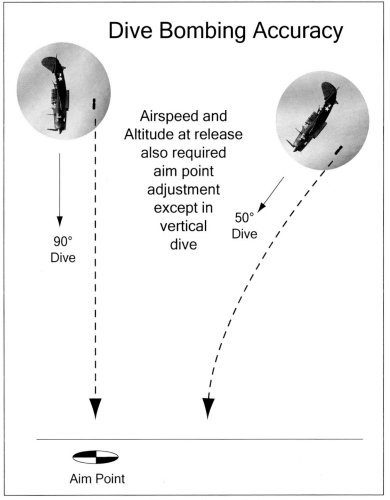

Dive Bombing Accuracy

Airspeed and Altitude at release also required aim point adjustment except in vertical dive

90° Dive

50° Dive

Aim Point

The steeper the dive, the less lead of the aiming point was required for airspeed, altitude, and gravity. It also presented problems for both defending fighters trying to follow a dive brake-equipped bomber in a dive and anti-aircraft gunners trying to elevate their guns high enough. (Author)

not all tactical airplanes were so equipped. Absent dive brakes, the attack had to be made at a lower angle, termed glide bombing.

Hitting the target with a bomb from a descent toward it required consideration of several factors. If the target was moving, as was the case with a ship, the pilot had to lead the target accordingly, so that during the bomb's fall, the ship moved to where the bomb was going. The attack angle also dictated the point of aim. The shallower the dive, the farther the bomb would fall short of the aim point, the amount depending on the lift/drag characteristics of the bomb. The altitude and speed at the release had the same effect—the higher the altitude and the lower the speed, the farther it would fall short. Finally, wind had to be taken into account. It was taken for granted that the pilot would insure that the airplane was not yawing, pitching, or rolling at bomb release, because that would also affect its trajectory.

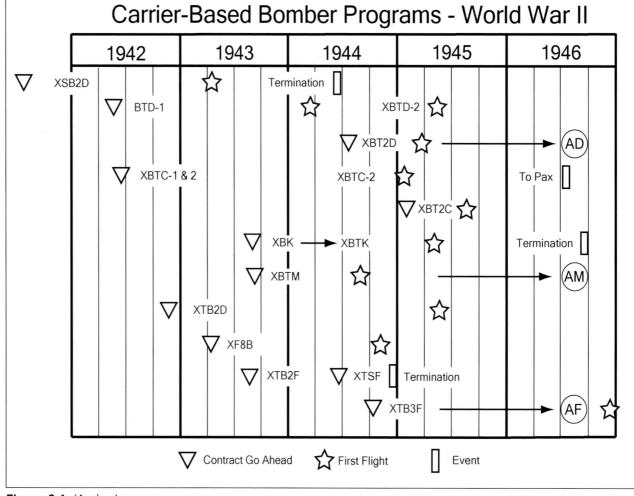

Figure 2-1. (Author)

The accuracy of a bomb drop is defined as the Circular Error Probable, which is the radius of a circle in which 50 percent of the bombs, rockets, or bullets will hit. This is stated as either a distance or as an angle measurement, mils, with one mil equal to one foot in 1,000 feet. The two are equivalent only when the same slant range is used at release. For example, a drop at an altitude of 5,000 feet above the ground in a 45-degree dive using a weapon system rated at 10 mils would have an accuracy of plus or minus 70 feet, but twice that distance if the release was at 10,000 feet. Mil markings on the bomb sight were also used to correct for wind, ballistic drop, target movement, etc.

For accuracy, it was best to minimize the number of variables. In other words, the pilot would decide in advance at what dive angle, altitude, and speed he would drop the bomb. Based on its ballistics, that would establish the aiming point to set on the gun-sight. After arriving at the target, he would then only have to estimate and correct for wind and the target's speed, if any. Relatively simple rules of thumb could be used for the final aim correction. For example, a bomb might take five seconds to fall from a release at 2,000 feet. During that time, a ship would move about 8 feet for each knot of its speed. The impact of wind would be the same. From 2,000 feet in a 50-degree dive, 8 feet would be about 3 mils vertically on the sight and four horizontally. A target traveling at 10 knots in no wind conditions would require a lead of about 80 feet or 30 to 40 mils in the sight.

Since it was likely that either dive angle or speed would be different than planned as the drop altitude neared, the pilot also needed to know how to compensate. For example, to release a bit low if the dive was not as steep as planned or a bit high if too fast—all this while being shot at and/or in turbulence, with the altitude unwinding at 400 to 500 feet per second.

At the end of the World War II, a new bomb director, ASG-10, had been developed to minimize the expertise, guesswork, and some

The armament shown here on the AM Mauler is impressive but would have had a significant impact on range, since armed with all of this the airplane was already at its maximum gross weight with no fuel in its tanks. The streamlined nose caps on the torpedoes were pulled off by a lanyard when the torpedo was dropped so the drag ring was effective. (Glenn L. Martin Maryland Aviation Museum)

of the risk involved in putting a bomb on target. Termed "toss bombing"—not to be confused with a later nuclear weapon delivery technique—it allowed the pilot to attack a target at varying dive angles and speeds with minimal manipulation of the bombing sight. After a relatively short, stabilized dive with the sight held on the target while the pickle button was pushed, the pilot was signaled to pull up straight ahead. At the appropriate instant during the pullout that the bomb's ballistics would result in it hitting the target, the director released it.

The director's simple three-tube analog computer took inputs from an altimeter and a gyro that measured dive angle. It could accommodate dive angles between 15 and 60 degrees, indicated airspeeds of up to 400 knots, and release altitudes of up to 11,000 feet with dive angles of 60 degrees. As the dive angle and airspeed decreased, the maximum release altitude decreased or the bomb would not reach the target in the toss. The pilot still had to correct for wind and insure that his pull up was through the target; otherwise the bomb would miss to the downwind side or in the direction of the pull up. A standard gunsight was used, "boresighted parallel to the flight path for average speed and loads in the dive."[5]

This equipment was initially installed in at least the AD-3 and some F4U Corsairs as the Mk 20 Mod 2 Computer CP-15A/ASG-10. However, it doesn't seem to have been well received in the fleet, as it doesn't appear to have been installed on the AD-4 or subsequent Skyraiders.

Postwar Attack Airplane Programs

Figure 2-1 summarized the various bomber programs initiated during World War II. In 1946, the Navy combined all the mission designations and functions of the scout bomber, the torpedo bomber, and the bomber torpedo into a single one: attack. All of the bomber squadrons became attack squadrons. The bombers in development or beginning production were redesignated: The BT2D became the AD and the BTM became the AM. The TB3F became the AF, even though it was immediately redirected to the high priority antisubmarine warfare role and not used as an attack aircraft. The last model of the F4U Corsair, optimized as a fighter-bomber for the Marine Corps, became the AU.

The AD was the most successful of all the carrier-based bomber programs initiated in World War II. Rugged and easy to fly, it had an excellent payload and range. Almost 3,000 were built for attack, electronic reconnaissance, and airborne early warning. It stayed in production for more than 10 years. The final combat sortie off aircraft carriers was made at the end of 1968, more than 20 years after VA-19A completed AD-1 service trials in December 1946. When it first deployed, the propeller-driven Grumman F8F Bearcat was the Navy's front line fighter. Its last deployment, by contrast, would be in the company of the supersonic McDonnell F-4 Phantom. After the Navy replaced it in attack squadrons, it soldiered on with the U.S. Air Force in support of helicopter combat rescue in Vietnam through 1972.

Although put into production, the AM was found more deficient than average during the Board of Inspection and Survey (BIS) acceptance evaluation at Patuxent River.[6] Correction combined with initial development problems meant that it was a year later than the AD-1s in being delivered to operational squadrons even though as the BTM it had been flown several months before the BT2D. It also had a more complicated control system, with spoilers and an elevator boost system having to be added during flight test. The Navy persevered with the big airplane, however, and ongoing production was used to equip five attack squadrons, the first of which received airplanes in March 1948 and participated in at-sea exercises in December 1948. Although Martin and the Navy worked to resolve its teething problems and modify the delivered aircraft, the AM reliability and carrier landing suitability was still unsatisfactory when it reached the fleet. It doesn't appear to have ever made an extended deployment, and was soon replaced by the AD in the deployable squadrons, being relegated to service with the reserves beginning in 1949. Only 149 were built.

The TB3F was redundant in late 1946 but concerns over the Russian submarine threat gave it new life as the platform for a dedicated ASW hunter/killer team to replace the aging TBMs that had been pressed into that role.[7] Since speed was no longer a virtue compared to endurance, the jet engine was deleted, freeing up the aft fuselage to be a crew compartment for sensor operators and eliminating the requirement for a crew position beside the pilot. The Wright engine option was also deleted in favor of the R-2800-46W. Fuel capacity was increased from 370 to 500 gallons.

The third TB3F prototype received an APS-20 radar installation as the XTB3F-1S, which flew in November 1948. The second prototype was modified to the XTB3F-2S configuration, including a sonobuoy receiver and small wing-mounted radar to localize a submerged submarine, a searchlight to illuminate it at night, and the TB3F armament capability—less cannons—to sink it with the aid of a periscope bombsight provided in the crew compartment. It first flew in January 1949. When the torpedo bombers were redesignated as attack airplanes in 1947, S for antisubmarine warfare not being used as a primary mission designation at the time, the XTB3Fs became AFs, with the AF-2W as the "hunter" and the AF-2S as the "killer."[8] However, the armed version was dedicated to the antisubmarine role and was only used operationally from ASW carriers. Grumman eventually produced 190 AF-2Ss and 156 AF-2Ws. An additional 40 AF-3Ss, which added magnetic anomaly detection to the localization toolbox, followed. The Guardians were physically the biggest of the single reciprocating-engine carrier-based attack airplanes, but powered by the smallest of the three engines available, the Pratt & Whitney R-2800. They were retired from active duty in the mid-1950s.

The AU-1 was a Corsair specifically modified for the Marine Corps close air support-mission. This one in flight test at Vought is loaded with 10 250-lb bombs on the wing stations. (Vought Heritage Center)

The AU Corsair was a gap-filler, ordered for a Marine Corps desperate for close air-support capability and waiting in line behind the Navy attack squadrons for the AD Skyraider. Originally designated XF4U-6, it was a minimal modification of the F4U-5, deleting the second stage of the supercharger since it would only be operated at relatively low altitudes. Relocating the oil coolers inboard and adding armor in the cockpit and below the engine and its accessories reduced ground-fire vulnerability. Like the F4U-5, it was armed with four 20mm cannons. The stores pylons were beefed up: the two inboard pylons could carry a 2,000-pound bomb, and the five new pylons on each outboard wing were capable of carrying 500-pound bombs. A more practical load-out was two 1,000-pound bombs and six 500-pound bombs. This allowed for full internal fuel plus one 150-gallon drop tank. The combat radius was only about 220 nautical miles (nm) but that was plenty for the shore-based Marines.

First flight of the AU-1 was accomplished on 29 December 1951. All of the 111 built were delivered to the Marine Corps by October 1952. It was in combat in Korea in June 1952, a month before NATC completed its evaluation, which wasn't complimentary: "The AU-1 is unacceptable for service use as an attack airplane with the present speed and acceleration restrictions and rocket launching limitations," and "The AU-1 is an unsatisfactory gun, rocket and bombing platform because of lateral-directional oscillations induced in dives." That didn't deter the Marines from continuing to fly missions with it or the French from borrowing some to augment their similar F4U-7s. Although it was carrier-qualified by NATC, it doesn't appear that any were ever based on U.S. Navy carriers.

The four reciprocating-engine-powered attack airplanes compared as follows:

	AD-4	AF-2S	AM-1	AU-1
Engine	R-3350	R-2800	R-4360	R-2800
Takeoff Power (hp)	2,700	2,300	3,000	2,300
Maximum Weight* (lb)	25,000	23,000	25,000	18,500
Gross Weight** (lb)	21,483	21,555	24,166	18,079
Internal Fuel (lb)	2,280	2,520	3,060	1,404
Armament (lb)	4,000	2,424	4,000	4,600
Wingspan (ft)	50	60	50	41

*Catapult takeoff
**Full internal fuel and armament as shown

One standard variant of the AD-1 through AD-4 was the Q for electronic reconnaissance and countermeasures. These aircraft had a single electronic countermeasure (ECM) operator position between the dive brakes. The crewman was provided with a windowed door on the starboard side and a small window on the port side. This is a factory-new AD-2Q. (Robert L. Lawson collection)

The AM Maulers were transferred to Naval Air Reserve squadrons without ever deploying. This one, assigned to NAS Grosse Ile, Michigan, is on display at an air show with 5-inch HVAR rockets on the wings, along with a 2,000-lb general-purpose bomb and two 1,600-lb armor-piercing or semi-armor-piercing bombs. (Robert L. Lawson collection)

The Grumman AF was solely assigned to antisubmarine warfare. This is the killer version, with an APS-4 under the right wing, a searchlight under the left, and a sonobuoy dispenser under the inboard wing. Weapons were homing torpedoes and depth charges carried in the bomb bay and rockets on the wing stations. (Robert L. Lawson collection)

The AD-4 Skyraider was produced in several different mission configurations. Shown here, from front to back, are the AD-4, the AD-4N, and the AD-4W. The AD-4N is carrying a torpedo and searchlight/sonobuoy dispenser for an Anti-Submarine Warfare (ASW) mission in addition to the permanently affixed APS-31 radar. The AD-4W provided detection of submarine periscopes/snorkels as well as airborne early warning. (U.S. Navy via Ed Barthelmes)

The night attack versions of the Skyraider had entry doors on both sides of the fuselage because there were two crewmen in the fuselage. This required deletion of the side-mounted dive brakes. This AD-3N has the APS-19A radar pod mounted on the left-hand wing pylon. It has been positioned on the shore-based catapult at Patuxent River for testing. (U.S. Navy via Ed Barthelmes)

Douglas AD Skyraider

Except for the -5, the basic configuration of the Skyraider varied little through the years. From the AD-1 onward, there were three pylons for bombs and 12 for rockets. Initially, there was only one 20mm cannon in each wing. Another was introduced during AD-4 production for a total of four cannons. Three large fuselage-mounted dive brakes limited speed in a vertical dive to 250 knots. The bomb displacing yoke for the centerline bomb rack of previous dive-bombers was replaced by an explosive charge that pushed the bomb clear of the propeller arc.

Production of each dash number up through -5 also included mission-specific variants. These included B for special armament (atomic bomb delivery in this case), N for night attack, Q for electronic reconnaissance and countermeasures, and W for airborne early warning. All of the variants except for the B incorporated a seat or seats for mission specialists in the fuselage.

In part due to Ed Heinemann's emphasis on weight control, the Skyraider had teething problems. He pushed his design and manufacturing team so hard that the first AD was 1,000 pounds less than the goal he gave them. An empty weight less than specification was and is rare. Being almost 10 percent underweight is all but unheard of. It turned out to be a bit too light from a structural standpoint, so the AD-1 was replaced in production, after 277 had been built, by the AD-2, with a beefed up landing gear attach structure and other structural changes (much of the beef-up was retrofitted to surviving AD-1s). Other -2 changes were a 2,700-hp military power-rated engine, a more ergonomic cockpit, and a different windscreen. At-sea

landings with beefed-up AD-1s and one AD-2 were accomplished in March 1948 aboard *Saipan* (CVL-48) to complete the demonstration of the landing-gear improvements. One made six landings with full internal fuel and three 1,000-pound bombs, an overload condition. Skin buckling was experienced in one wing skin area, which was subsequently beefed up during overhaul. Only 178 AD-2s and two-place -2Qs were built due to postwar budget limitations.

The beef up continued with the AD-3, which incorporated a modified landing gear with a longer stroke to accommodate higher sink-rate landings and fuselage reinforcements. Other detail improvements included changes to the cockpit, an improved Aeroproducts propeller, and provision for emergency operation of the canopy via an air bottle. At-sea carrier trials were accomplished in February 1949. The maximum gross weight demonstrated was 18,578 pounds, which included full internal fuel and two 150-gallon external tanks, 250 pounds of ammunition, and one 2,000-pound bomb. There were 194 AD-3s, including 31 AD-3Ws for airborne early warning.

The AD-3 was well-regarded by the Board of Inspection and Survey, but additional improvements were recommended:[9]

Advantages in performing primary mission:
- Excellent dependability and relatively easy to maintain.
- Very well suited to carrier operations.
- Adequate range and endurance.
- Has a very desirable load carrying capacity.
- It has adequate performance on strike missions at altitudes below 20,000 feet. Unencumbered by external stores, it can compete favorably with most propeller-driven fighters at altitudes below 15,000 feet.

This Skyraider, still a BT2D, is armed with two Tiny Tim rockets, 12 HVARs, and two 20mm cannons. The three large fuselage dive brakes have been opened to allow a steep dive without gaining excessive speed. (U.S. Navy via Ed Barthelmes)

Ordnance men on aircraft carriers wear red jerseys and/or helmets. Here, an AD Skyraider is being loaded with 500-lb general-purpose bombs. The tailfins were built, shipped, and stored separately from the bomb bodies, which is why they are a different color and condition. The fins were easily damaged, which would affect the bomb's ballistics and hence its accuracy. (Robert L. Lawson collection)

Disadvantages:
- Two 20mm (cannons) are not considered adequate firepower—four are desired
- The APS-4 radar equipment is superfluous for day missions unless provisions are made for an operator.

All-weather suitability:
The AD-3 is not considered suitable for all weather operations because of the following deficiencies:
- The poor arrangement of flight instruments causes excessive pilot fatigue during long periods of instrument flight. This condition is accentuated by the lack of an automatic pilot.
- There is no protection against the hazards of icing.

- The performance of the APS-4 radar is not adequate for instrument flight, especially since the pilot must fly and interpret the scope alone.
- Precipitation static renders the operation of the AN/ARC-5 receiver unreliable.

Recommendations:
- Re-position the pilot's seat to allow a better "in flight" weight distribution, thereby reducing pilot fatigue.
- Relocate flight instruments to provide a less fatiguing "scan pattern" for the pilot.
- Equip with de-icing and anti-icing equipment.
- Replace the AN/APS-4 radar with AN/APS-31 or AN/APS-19A radar.

The F6F drone was aimed by reference to a picture transmitted from a TV camera carried in this pod under the right wing. The transmission antenna is just visible above the wing. A TV receiver with a viewing hood attached is located adjacent to the tire for checking the reception. (National Archives)

This F6F drone marked V5 and armed with a 1,000-lb bomb slung under the belly, has been positioned on the port catapult of Boxer for launch. A crewman and a pilot are making final cockpit checks. The crew of the AD-4N in the foreground will help guide the drone to the target. (National Archives)

The AD-3N was the first production night-attack variant. It had two crew seats in the fuselage, which required deletion of the side-mounted speed brakes. An APS-19 radar, more capable than the standard bomb rack-mounted APS-4 option, was standard equipment, along with radar countermeasures detection and jamming equipment. However, the BIS report stated that the APS-19A did not provide adequate range and bearing information and recommended that it be replaced with the APS-31 radar or range and bearing markers be provided on the APS-19A scope.

The AD-4 replaced the AD-3 on the Douglas production line in mid-1949. A welcome addition was a P-1 autopilot. The APS-19 radar was substituted for the APS-4 after the first 28 were delivered. Including the usual Nan, Queen, and Whiskey variants and the new Baker one for atomic bomb delivery, a total of 1,051 were produced, making it the most numerous Skyraider model. It was to do most of the heavy lifting in the upcoming Korean War. The AD-4 was the first Skyraider to have four 20mm cannons, a change incorporated after 210 had been produced and retrofitted to those aircraft if they reached an overhaul.

The AD-4N was a further improvement over the AD-3N. Upgrades included the APS-31 radar, although the first 28 were delivered with the APS-19. The new radar was a semi-permanent installation in place of the right-wing bomb pylon. This provided a significantly improved capability for horizontal bombing of a suitable target by radar, with the two crewmen in the aft fuselage providing navigation and countermeasures assistance to the pilot as they did in the AD-3N. It could carry a large searchlight on the left wing pylon, but this was primarily for ASW mission capability, since it would have provided far too convenient an aiming point for enemy antiaircraft guns. Flares were used if required for visual attacks at night.

The Korean War

When the Korean War began in June 1950, the AD Skyraider immediately began doing the heavy lifting for close air support and interdiction of North Korean supply lines. However, it proved vulnerable to ground fire in spite of the original 186 pounds of armor, bullet-resistant windshield, and "deflector plates" to protect the pilot and engine accessories. (Deflector plate was the term of art for heavy-gauge aluminum that provided protection against shell fragments and shallow angle projectile strikes.) Another 618 pounds of aluminum armor deflector plates was therefore added externally to better protect the sides of the cockpit and the engine accessory area. This additional protection was first utilized in combat in March 1952 and proved effective. An article in the May 1953 edition of *Naval Aviation News*, reported that the loss of at least 18 Skyraiders, the full complement of an air group, had been avoided by the new armor, not to mention injury to or death of the pilots.

This is an early AD-4 Skyraider (only one 20mm cannon per wing) armed with a 2,000-lb bomb and six anti-tank rockets on this wing. The ordnance man is arming the bomb. (National Archives 80-G-428979)

from the existing F6F target drone was the addition of a TV camera mounted in a pod hung beneath the right wing with a transmitter aerial mounted on the upper surface of the wing. A 2,000-pound bomb was hung from the centerline pylon. Between 28 August and 2 September, the six F6Fs were launched against railroad bridges, a hydroelectric plant, and a railroad tunnel. The results were disappointing: one direct hit, three near misses, one drone that failed to reach the target, and one for which a damage assessment was not obtained. No additional drones were assigned to the program, and the operational requirement was canceled in early 1954.

Night attacks flown in the AD-4N were accomplished with only limited use of the radar due to the size of the targets. Its primary contribution was navigation to and from the area of interest and for letdowns and climb outs in mountainous terrain. Once there, the pilot sought out targets visually, usually dropping flares to prosecute the attack. The harsh Korean winter weather resulted in the hasty development of a weatherized AD-4NL. This was the night-attack version modified with deicing and anti-icing systems that had been created for the AD-4L for true all-weather operation. About 36 AD-4NLs were created.

While the U.S. Navy maintained the torpedo-delivery capability with the AD, AF, and AM, it was never again called upon, at least against ships. It did come in handy against an unusual Korean War target, the Hwachon Dam. In April 1951, United Nation tacticians were concerned that the Communists in control of the dam would use it to either stall a UN advance or facilitate a North Korean attack. After Army Rangers were unable to recapture the dam and Air Force B-29s had bombed it without effect, the Navy was asked to try to incapacitate the sluice gates that controlled the release of water. The first attempt was made by VA-195 AD Skyraiders armed

The torpedo made a brief but important comeback during the Korean War when it was called upon to destroy the sluice gates of the Hwachon Dam. This VA-195 Skyraider photographed while en route to the dam is being flown by Ens. Robert Bennett, who is ignoring a loose carburetor air scoop inlet. Squadron skipper, LCDR Harold "Swede" Carlson, took the photo. (U.S. Navy via Ed Barthelmes)

Another attempt at vulnerability reduction was the issuance in January 1952 of an operational requirement for a catapult-launched assault drone that could be created by the addition of a kit to an existing carrier aircraft. As it happened, a suitable one was readily available: the F6F-5K target drone. Guided Missile Unit 90 was established in July 1952 to provide an operational evaluation of the concept and assigned six F6F-5Ks and two AD-2Qs as control aircraft. In less than two months, GMU-90 was launching a drone strike against a target in North Korea from *Boxer* (CV-21). The technology was not very far advanced from that of the TDN/TDR program. The major change with 2,000-pound bombs and Tiny Tim rockets. Neither proved effective. The commanding officer of *Princeton* (CV-37), CAPT William O. Gallery, then resorted to World War II-vintage Mk 13 torpedoes.[10] In a hail of flak and bullets from anti-aircraft emplacements along the heavily defended shores of the lake, eight ADs made runs on the dam, dropping torpedoes that destroyed two gates and damaged the third.

Atomic ADs

Like carrier-based photoreconnaissance and airborne early warning, all-weather attack and nuclear weapons delivery were originally accomplished by specially equipped airplanes flown by aircrew specifically trained for the mission. The aircrew and airplanes were assigned to large shore-based "composite" squadrons, one on each coast, which established mission doctrine and accomplished the training. Small detachments from these squadrons were assigned to the carrier's air group for operational deployments.

For example, VC-35 was formed in May 1950 at NAS San Diego, California, to provide anti-submarine detachments to Essex-class carriers. (VC-33 was the East Coast equivalent.) As a result of the Korean War, the squadrons' role was expanded to night attack. A typical detachment would consist of four airplanes, six pilots, and about 40 enlisted men including aircrew. The Skyraiders involved were the multi-place AD-4N and the single-seat AD-4B, which had been modified with a nuclear weapon delivery capability.

In the late 1950s, the existing attack squadrons in an air group absorbed the specialized attack missions. VA(AW)-35's last detachment returned from a western Pacific deployment aboard *Lexington* in December 1959.[11]

VC-35 received delivery of the first AD-4Bs in early 1952. Training was accomplished at NAS El Centro, California, (delivery tactics) and Sandia Base, New Mexico (stores management). Modifications included wiring for the management of the store and a fuselage recess for its tailfins. After analysis of the thermal effect on the relatively thin skin of the control surfaces, these were reportedly repainted from dark blue to white.

The AD-4B's primary weapon was a gun-type device, the Mk 8, which had a lightweight case so it weighed "only" 3,250 pounds. It was designed for delayed-action detonation to destroy underground or well-protected facilities such as submarine pens. It could supposedly penetrate 22 feet of reinforced concrete. It was initially delivered in a dive with an immediate course reversal after release to escape the burst effects.

In December 1951 while docked at Oakland, California, *Philippine Sea* (CV-47) was used for an evaluation of an "emergency" capability for assembling an atomic weapon on a non-Midway-class carrier. In June 1952, VC-35 Special Detachment William was deployed to Korea on *Essex* (CV-9), which had been modified for storage and handling of atomic bombs. The AD-4Bs were then based in Japan with their Mk 8s stored aboard the carrier. VC-33, based at NAS Atlantic City, deployed AD-4B

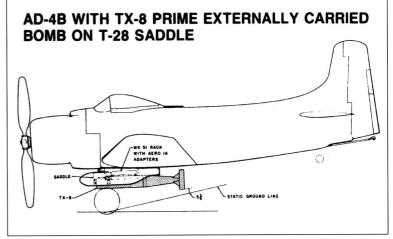

AD-4B WITH TX-8 PRIME EXTERNALLY CARRIED BOMB ON T-28 SADDLE

The Mk 8 atomic bomb did not initially have an internal arming capability and required the AD to be equipped with a "saddle" that contained a mechanism to perform this function. (Jay Miller collection)

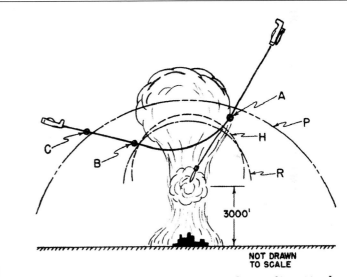

LINE R here defines the radiation limit; line H, the heat limit; and line P, the blast limit. Now if the aircraft drops the bomb at A, is at B at the time of explosion, he will be at C when the blast nears him.

This illustration appeared in an article in the February 1953 issue of the Naval Aviation Confidential Bulletin. *It provided a summary of the considerations for the safe delivery of a nuclear weapon in a dive, which was the initial method to be used by the AD-4B Skyraider to insure accuracy.* (U.S. Navy via author's collection)

detachments aboard the Midway-class carriers operating in the Atlantic Ocean and the Mediterranean Sea. AD-4B production totaled 193 airplanes.

CHAPTER THREE

LONG-RANGE NUCLEAR BOMBERS

The AJ Savage was the first long-range, carrier-based bomber capable of carrying the big Mk 4 bomb. This is an AJ-2 in the twilight of its career, operating as an inflight refueling tanker from Lexington in 1956. (Robert L. Lawson collection)

The successful demonstration of the atomic bomb challenged each of the U.S. military services to reorient itself to a different way of waging war, if one became necessary. The soon-to-be-independent Army Air Forces believed that their new B-36 intercontinental bomber, the prototype of which was to finally fly in August 1946, would make the Army and Navy all but superfluous: a war with the likely enemy, the Soviet Union, which had a huge army and limited dependence on raw materials transported by sea, could

only be won by strategic bombing. With its 10,000-mile range and claimed invulnerability from fighter interception and anti-aircraft defenses, B-36s could drop atomic bombs on important industrial complexes in the Soviet Union from U.S. bases in Alaska and Maine, even if bases in allied countries and fighter escorts weren't available.

By the end of the war in the Pacific, the U.S. Navy's primary mission for its carriers was no longer sinking enemy ships but supporting Marine assaults from the sea. However, this was of limited value in a European war involving the Soviet Union. This potential adversary was slowly building a naval force, but it was primarily littoral and defensive in nature, lacking aircraft carriers. The main concern in keeping the seas open for commerce and the passage of U.S. war ships transiting to protect U.S. interests abroad was the prospect of Stalin building a fleet of submarines using German technology.

The National Security Act of 1947 reiterated the Navy's control of carrier and land-based aircraft related to the sea-control mission. The Navy would maintain a presence wherever it could take a carrier, supporting allies that ringed and were potentially threatened by the Soviet Union and China, soon to be under the control of the Communists. The next battle was fought in 1949 over the Air Force having sole control of strategic bombing, in effect meaning delivery of atomic bombs.

Navy officers had been involved in the Manhattan Project that resulted in the atomic bomb. Navy CAPT William "Deak" Parsons was assigned to the Project's Ordnance Division in March 1943. His prior experience included participation in the development of a proximity fuse for anti-aircraft shells. In addition to playing a key role in the design and development of the proximity fuse for the atomic bomb, he became responsible for the planning and execution of the Army Air Forces' delivery of the bombs; was in charge of their assembly and checkout on Tinian Island; and was aboard *Enola Gay* on the first atomic bomb mission. CDR Frederick (Dick) Ashworth, a Naval aviator and former TBF squadron commander, was his Director of Operations. Another Navy combat aviator, CDR John T. "Chick" Hayward was assigned to the project, doing implosion device testing at the China Lake Naval Ordnance Test

Major Soviet Union Navy Targets

Figure 3-1. *The U.S. Navy's major targets for first-day nuclear strikes in the 1950s were the Soviet Union's naval facilities from which it would attack U.S. aircraft carriers. The Northern Fleet was located at Severomorsk Naval Base near Murmansk and several other facilities located along the southern shore of the Barents Sea and bordering the White Sea at Arkhangelsk. The Baltic Sea Fleet was located at Kaliningrad—the Russian enclave between Poland and Lithuania; the Kronstadt Naval Base near Saint Petersburg; and at bases in Latvia, Lithuania, and Estonia. The Black Sea Fleet was based in Crimean ports including Sevastopol. In the east, the Pacific Fleet was based at Vladivostok and Vilyuchinsk on the Kamchatka Peninsula. Not until 1960 would the Navy's nuclear strike planning be coordinated with the Air Force's as part of SIOP, the Single Integrated Operational Plan.* (Author)

There were two live-fire tests in July 1946 at Bikini Atoll in the Pacific to evaluate the vulnerability of ships to atomic bombs. *Able* was an Mk III drop from a B-29 on a fleet of obsolete Navy and former German and Japanese ships; *Baker* was a 23-kiloton underwater detonation of a moored Mk III. *Able* missed the target, the battleship *Nevada*, by 2,000 feet and it did not sink, much to the delight of the Navy participants. *Baker* couldn't miss, but the Navy believed the overall results indicated that ships were less vulnerable to an atomic bomb than assumed. (U.S. Navy via author's collection)

Station, having done graduate work in physics at the University of Pennsylvania in the late 1930s.

Two different atomic bombs were produced in the Manhattan Project. Both were based on the discovery that the near instantaneous creation of a critical mass of uranium or plutonium isotopes in the presence of a neutron initiator would result in an unprecedented powerful chain reaction. The mechanism used to create a critical mass in the Mk I bomb was the so-called gun-type that used a uranium isotope: an open-ended assembly of uranium rings likened to a roll of Lifesavers was fired down a gun tube onto a solid uranium cylinder. Apart, the two uranium components were subcritical; together they were a critical mass. The much more complex design for the Mk III bomb relied on shaped charges that compressed a plutonium sphere into a critical mass around an initiator.

The gun-type bomb, codenamed Little Boy, was so straightforward in concept that one wasn't test fired before it was dropped on Hiroshima on 6 August 1945.[1] The implosion design was detonated in the New Mexico desert on 16 July to prove that it did function as planned; that cleared the way for a Mk III Fat Man being assembled on Tinian, an island in the Pacific from which the Japan-bound B-29s

departed, to be dropped on Nagasaki on 9 August. Both were far heavier—almost 9,000 pounds for the Mk 1 and 10,300 pounds for the Mk III—than could be lifted by an existing carrier-based aircraft. The aptly named Fat Man was also 5 feet in diameter and almost 11 feet long.

All the early atomic bombs were manually armed in-flight after takeoff. In the case of the Hiroshima and Nagasaki bombs, this was done by the U.S. Navy officers who were part of the Manhattan Project. Because of the lack of built-in safety devices on the gun-type bomb, CAPT Parsons also loaded its explosive charge after the take-off of *Enola Gay*. CDR Ashworth had the responsibility of arming the Fat Man aboard the B-29 *Bockscar*[2] on the way to Japan. As the so-called weaponeer, each was also the tactical commander, the final authority for the decision where to drop the bomb since they alone had the technical knowledge of the bomb's operation and capability.

Once demonstrated to work, the implosion concept was preferred, although the gun-type mechanism was used for the Navy's Mk 8 concrete-penetrating bomb, with additional safety devices. Only a handful of Mk Is were built, as they required much more uranium than the implosion device did of plutonium. Another drawback was the potential for the mechanical creation of a critical mass in the event of a crash of the bomber. While unlikely to produce a full nuclear event, even an atomic fizzle was to be avoided if at all possible.

The Mk III was followed almost immediately by the externally identical, except for the stabilizing fins, Mk IV. The spherical shape and size of the implosion device dictated the diameter, and the B-29's modified bomb bay dictated the length, which was a bit short from a ballistics standpoint.

If the Navy wasn't committed to having a nuclear strike capability before Hiroshima and Nagasaki, they certainly were after. In September 1945, the Chief of Naval Operations established the Special Weapons Division of the Office of the CNO headed by a VADM. Planning began for the construction of a super carrier that could launch and recover aircraft up to 100,000 pounds gross weight and carrier-based jet or turbo-prop powered bombers that could carry a bomb

The use of aircraft carriers to launch relatively large and long-range bombers was demonstrated by Jimmy Doolittle's attack on Japan in April 1942. *Hornet* (CV-8) carried 16 Army Air Forces B-25 medium bombers to the takeoff point about 600 miles east of Japan. After bombing their targets, Doolittle and his fellow pilots were then to proceed to landing fields in China. (National Archives 80-G-41197)

weighing up to 12,000 pounds to a radius of up to 2,000 nm. The new airplane had to be that big to carry the existing atomic bomb from an aircraft carrier located well offshore—making it more difficult for it to be located and attacked—and still reach all the Soviet Navy bases, shipyards, and air fields in a region as shown in Figure 3-1.

BuAer's three-phase carrier-based strategic bomber plan was issued in December 1945, with Phase 1, development of a 45,000-pound gross weight bomber with modest capability, to be initiated immediately. The January 1946 Request for Proposal (RFP) stated that the aircraft "will be taken aboard a carrier as a special loading and need not be capable of being struck below" although it was "highly desirable."[3] The specification called for the delivery of an 8,000-pound bomb at a distance of 300 nm, subsequently increased to 600 to 700 nm. The bomb bay was required to carry a cylinder 5 feet in diameter and 16 feet long that was hung from a single point. The cockpit was to be pressurized, but in-flight access to the bomb bay was required so the bomb could be armed after takeoff.

Takeoff by a 600-foot deck run from a Midway-class carrier was required. Other performance requirements were:

Speed at 35,000 ft, 60% Fuel, Jet(s) On	500 mph
Sea Level Rate of Climb	2,500 fpm
Stalling Speed, Maximum Gross Weight, Power On	less than 100 mph
Service Ceiling	45,000 ft

Unfolded size was defined by the capability for one bomber to pass another on a Midway-class flight deck aft of the island with both having the wings spread. Maximum landing gear tread was 24 feet to allow for catapulting with adequate wheel clearance from the deck edge.

Three proposals were received. Consolidated Vultee's basic design was rejected out of hand because it was over the specified landing weight and short in combat radius by 40 miles. Two R-2800 turbocharged engines and two Westinghouse J34 jet engines powered it, one in each nacelle. Douglas submitted a second configuration after the 1 May proposal due date which was also evaluated since it was superior to the first. It was also powered by two R-2800 turbocharged engines but with a single General Electric I-40 (produced by Allison as the J33) in the fuselage. The North American design was powered by two R-2800 turbocharged engines like the others and a single GE I-40 like Douglas, but with optional afterburning. Its wing had a lower aspect ratio than Douglas', which resulted in a lower empty weight. That meant that the Douglas design would have a range advantage with a mission radius of more than 600 nm but the North American design benefited by having the lightest landing weight, shortest takeoff distance, and best climb at altitude. The maximum speeds were approximately the same without afterburning, which BuAer was not enthusiastic about. The North American proposal was also the lowest in cost.

Having won the design competition, North American Aviation, Inc. received a letter of intent in June 1946 for its XAJ-1, which was powered by two Pratt & Whitney R-2800 turbocharged engines and one Allison J33 jet engine. The jet was only used for takeoff, landing (in case a wave-off was required), and the run-in to the target for extra

The AJ Savage was just big enough on the inside to carry the Mk III atomic bomb and just small enough on the outside, when the wings were folded, to fit on the carrier's elevators and in its hangar. This XAJ's picture was taken on 2 September 1948. (National Archives 80-G-706015)

speed. Its flush NACA-type intake was located on the top of the fuselage and incorporated a hinged door that was closed when the jet was not operating.

Changes after the award included widening and deepening the bomb bay to accommodate "the ultimate primary bomb" and adding tip tanks for an overload condition to provide a radius of 700 nm with a reduction of the time-over-target requirement at a high power setting.

The AJ was to have a crew of three: pilot, bombardier/navigator, and the third crewman. The AJ full-scale mockup was reviewed in October at NAA's Los Angeles facility. The acceptability of emergency egress without an ejection seat was a discussion topic. Weight concerns eliminated ejection seats; the crew was expected to clamber out the entrance door if a bailout were required. Another debate was whether a stick or wheel was preferable for the pilot's control; this was settled in favor of a stick at the time.

Instead of being positioned in the nose, the AJ bombardier sat beside the pilot and operated the ASB-1 bomb director set, originally known as the Bomb Director Mk 5 Mod 0. The ASB-1 was a radar/optical bombing system that could also be used for navigation and search. Developed by Norden Laboratories Corp., it became the standard bombing system for the Navy's heavy attack aircraft for many years. The system was computer driven and combined a periscope view with radar displays. There were three radarscope presentations. One was stabilized with north up for comparison with a map. Another automatically tracked the periscope line-of-sight for ease of initial point verification. The third provided an expanded view of any selected area and could be oriented either north up or to track with the periscope view. It weighed almost 1,000 pounds, which limited its use to the heavy attack aircraft.

As originally specified in 1949, the bombing system allowed for some evasive maneuvering on the run-in: altitude changes of 3,000 feet

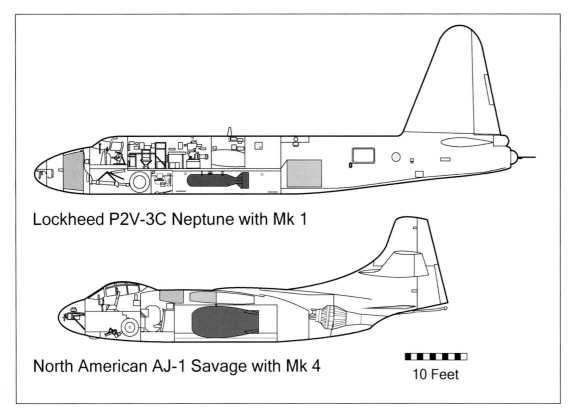

Lockheed P2V-3C Neptune with Mk 1

North American AJ-1 Savage with Mk 4

10 Feet

The P2V Neptune bomb bay could not accommodate the Mk 4, which was the successor to the Mk III Fat Man that destroyed Nagasaki. However, the Mk I did fit in the bomb bay, which was configured to provide access to arm the weapon. The Navy therefore adapted it to be its interim carrier-launched, all-weather, long-range atomic bomber while the AJ-1, with a bomb bay sized for the Mk 4, was being designed and flight-tested. (Author)

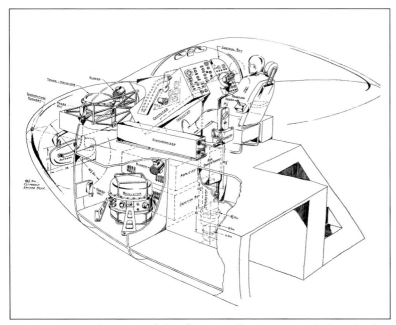

It wasn't just the size of the first production atomic bomb that required a big airplane. An all-weather delivery capability in the 1950s necessitated a bulky radar system and a bombardier, as shown in this diagram. The ASB-1 Radar Bombing System included a periscope as well for target verification and as a visual bombing option. (U.S. Navy via author's collection)

per minute climb to 6,000 feet per minute descent within plus or minus 5,000 feet of the bomb release altitude; airspeed excursions of plus or minus 50 knots; bank angles of up to 30 degrees with radar sighting and 60 degrees with optical sighting; and lateral excursions from course, as long as the bomber was on course when the bomb was released. Aiming points could be offset up to five miles from a reference point with a good radar return if the target itself did not provide an adequate one. At the drop point, the system automatically opened the bomb bay doors, released the bomb, and closed the doors. However, it was some years before it was perfected and available for installation in the AJ.

The size of the AJ dictated that the controls be hydraulic-powered. Almost all the other functions, including the cabin pressurization compressor, were hydraulic powered as well, resulting in five separate 3,000-psi hydraulic systems.

Interim Capability—The P2V Neptune

In the meantime, the Navy needed to establish an atomic bomb-delivery capability from aircraft carriers as soon as possible. BuAer therefore began a program to modify the Navy's P2V twin-engine patrol bomber for the mission. It was not nearly as big as the B-29 but still three or four times as heavy as any carrier-based airplane. Its primary asset was its ability to carry a 10,000-pound bomb, albeit the smaller Mk I due to the dimensions of its bomb bay. It also had world-record range. Since it could only operate from the three Midway-class carriers and only a handful of Mk Is were expected to be available, just 12 P2V-3Cs were ordered.

Col. Jimmy Doolittle had successfully employed the concept of deck launching a relatively long-range, twin-engine bomber from an aircraft carrier in his attack on Japan in April 1942. In September 1946, an early version of the P2V flew from Perth, Australia, to Cleveland, Ohio, non-stop and unrefueled, a distance of 11,235 miles, almost

The P2V-3C prototype was a modified P2V-2, Lockheed build number 1080. It was stripped of virtually all external excrescences and the radar relocated to the nose. The 20mm tail gun defense was retained. (U.S. Navy via author's collection)

Extended Hydroflap

To facilitate ditching if required at the end of its mission, the P2V-3C had a hydroflap added behind the nose gear well. This device, which when extended acted to keep the aircraft's nose up in a water landing, had been demonstrated to increase the survivability of water landings in a B-24 test. (U.S. Navy via author's collection)

halfway around the world. (Additional fuel tanks in lieu of a payload provided the extra range, and JATO permitted the overweight takeoff.) At a gross weight of 74,000 pounds, the P2V was far too heavy for existing catapults, so it was to be launched with a deck run augmented by eight 1,000-pound thrust JATO rockets, which were jettisoned along with the attaching racks after takeoff. The P2V was powered by the turbo-supercharged Wright R-3350 with 3,200 hp available for takeoff. Cruising speed was less than 200 mph, with top speed about 340 mph.

The Navy originally intended that the P2V also land back aboard the carrier so the modification and development program included a tailhook. To maximize range, additional fuel tanks were installed in the wings, and tanks were added to the forward and mid fuselage. Total internal fuel was thereby increased by 75 percent from 2,350 gallons to 4,120. To reduce drag, the dorsal turret, rocket launchers, tail skid, astrodome (replaced by a periscope sextant), and most of the antennas were removed. The nose guns were deleted and the AN/APS-31 search/navigation radar relocated to the nose. The radar bombsight was the APA-5, which used the return from the AN/APS-31. The bomb bay was modified with a single shackle and platforms for in-flight arming of the weapon. Since the flight time possible with the onboard fuel now exceeded the engine oil capacity, a 38-gallon oil tank was installed at the navigator's station (he sat on it) and plumbing provided to replenish the oil tanks in the nacelles in flight. To reduce weight, one (of four) engine-driven electrical generators and the emergency hydraulic system was removed (a manual nose gear extension feature was substituted).

The brace of 20mm tail guns was retained since the P2V would be unescorted. Radar countermeasure equipment was added to detect the presence, characteristics, and bearing of threat radars but there was no jamming capability. The crew was reduced to four: pilot, copilot/weaponeer, bombardier/navigator, and radioman/tail gunner. The first takeoff evaluation at sea was the launch of two P2V-2s from *Coral Sea* (CV-43) in April 1948.

The newly established U.S. Air Force took exception to the Navy's effort to establish a separate strategic bombing capability. However, in April 1948, following a March meeting at Key West, Florida, of the Secretary of Defense James Forrestal, the service secretaries, and the Joint Chiefs of Staff, the Navy received authorization "to conduct air operations as necessary for the accomplishment of objectives in a naval campaign." The Air Force retained responsibility for strategic air warfare, which was defined as "Air combat and

It was intended that the P2Vs land aboard to be loaded with an atomic bomb and then launched. A tail-hook attachment was designed and, as shown here, proof tested. Shore-based arrested landings were also accomplished. Nevertheless, the airplane was never landed on a carrier. (U.S. Navy via author's collection)

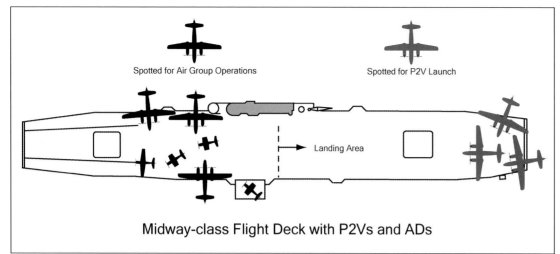

Spotted for Air Group Operations

Spotted for P2V Launch

Landing Area

Midway-class Flight Deck with P2Vs and ADs

Figure 3-2. Depending on the mission range, probably no more than three or four P2V-3Cs could be launched from a Midway-class carrier and leave enough open deck for the first take-off run. Note that very few, if any, of the air group's aircraft could be on deck during P2V take-offs. Since no aircraft can be parked aft of the island during landings, all the P2Vs would have to be spotted forward for at least that part of the cycle. They could be left there, but keeping more than two P2Vs forward blocked one of the catapults. In other words, the carrier's ability to operate its air group was severely restricted when P2Vs were on board. (Author)

On a P2V-3C launch from a Midway-class carrier, there wasn't much margin for error in either lineup or distance. This is a take-off from Franklin D. Roosevelt **on 2 April 1949. The white line beneath the nose wheel provided the necessary clearance from the carrier's island and the port-side deck edge.** (National Archives 80-G-400049)

supporting operations designed to effect, through the systemic application of force to a selected series of vital targets, the progressive destruction and disintegration of the enemy's war-making capacity to the point where he will no longer retain the ability or the will to wage war." It was clear to the Navy at least that the Air Force's focus would therefore be on urban and industrial targets as it had been in Germany and Japan, while the Navy would take on military bases and deployed forces, employing atomic weapons if necessary. The Air Force took exception to this interpretation but was overruled by the Secretary of Defense later that year at a follow-on conference in August at the Naval War College in Newport, Rhode Island.

The Navy failed to qualify the P2V for carrier landings in spite of a concerted program to do so in 1949.[4] If it couldn't return to a base onshore, the crew was to either bail out or, if over water, ditch.

There probably wasn't a need to recover the aircraft back aboard for rearming since there were apparently not as many Mk Is as there were P2V-3Cs, but it would have been convenient to fly the airplanes aboard if the mission became necessary. Instead, the carrier had to proceed to a dock and crane the P2Vs aboard, a time-consuming activity that signaled U.S. intentions and increased the vulnerability of the carrier. Having P2Vs aboard also hampered operations of the carrier's air group since the airplane was huge and didn't fold. It therefore couldn't be stowed on the hangar deck and either precluded operation of one of the catapults or had to be respotted following every launch and recovery as shown in Figure 3-2.

In anticipation of the first operational use of nuclear weapons, the Navy formed three Special Weapons Units, one for each of the Midway-class carriers, *Midway* (CV-41), *Franklin D. Roosevelt* (CV-42), and *Coral Sea* (CV-43). Based and trained at Kirkland Air Force Base, New Mexico, these units were responsible for all Navy atomic bomb assembly, maintenance, and checkout. (The early bombs required teams of about 50 men and 80 hours to assemble with frequent disassembly and rechecks required for battery and initiator replacement.) Modifications to these carriers were also accomplished to provide for atomic bomb handling, storage, maintenance, and loading. A Special Weapons Unit accompanied the atomic weapons whenever they went aboard carriers. One loaded an AJ with an Mk 4 and an Mk I aboard *Coral Sea*, first at dockside and then at sea in early 1950 to develop procedures and equipment.[5]

The first heavy attack squadron, VC-5, was commissioned at Moffett Field, California, on 9 September 1948. The acting commanding officer was CDR Dick Ashworth, the weaponeer on *Bockscar*. He became the executive officer when CAPT John T. "Chick" Hayward arrived in January. Deliveries of P2V-3Cs to the squadron began in November 1948. In February the squadron flew three of them to NAS Patuxent River for JATO training. In early March, the airplanes were hoisted aboard *Coral Sea* for the squadron's first exercises at sea. Chick Hayward was the first to take off. His P2V-3C was loaded with a dummy 10,000-pound bomb and fueled to a gross weight of 74,000 pounds. He flew to the West Coast, dropped the simulated bomb, and returned nonstop to Patuxent River. The other two P2Vs also launched for pilot qualification, but at lower gross weights.

Later in March, two more VC-5 crews flew to Patuxent River for carrier qualification. Their at-sea takeoffs were from *Midway* on the 26th. This was followed by another loading and launch on 7 April, the exercises being as much for ship's crew familiarization as squadron training. At the time, all the big Midway-class carriers were home ported on the East Coast and deploying to the Mediterranean or the North Sea.

North American AJ Savage

While VC-5 was establishing and promoting an atomic bomb delivery capability with the P2V, North American was developing the AJ Savage. Following its first flight in July 1948 by NAA's test pilot Bob Chilton from Los Angeles International Airport, the Savage was rushed into service, notwithstanding its first fatal crash in February 1949 when the tail came off during a side-slip maneuver, killing test pilots Al Connor and Chuck Brown. The rudder was intentionally large initially, in order to permit landing approaches at low speed with

*VC-5 made several long-range flights for aircrew training, but also to publicize its capability to fly 5,000-mile missions in the P2V-3C after takeoff from a carrier. This is a takeoff from the USS **Midway** in April 1949.* (National Archives 80-G-707168)

one of the reciprocating engines inoperative. However, it was far too large if deflected during cruise flight. This resulted in flight restrictions until the empennage was modified to delete the dihedral and reduce the size of the rudder. (The strakes on the side of the fuselage above the jet exhaust were also deleted.) The modified tail was retrofitted to surviving AJ-1s.

This XAJ-1 picture appears to have been taken at Patuxent River. Note the wrinkles on the aft fuselage. These were very evident at the roll-out ceremony of the first Savage, causing some concern among the dignitaries present. (National Archives)

The Landing Signal Officer (LSO) has just given a cut to an AJ-1 landing on Lake Champlain *(CV-39), an axial deck carrier.* (U.S. Navy via author's collection)

Not only was manually folding the AJ Savage a time-consuming, labor-intensive chore, even then it was a tight fit below decks. Note that two tow tractors are ganged together to provide the needed horsepower, and two sailors are insuring clearance of the tip tanks with the overhead. The ones crouching on the inboard side of the main landing gear are poised to chock the tires if matters get out of hand. (U.S. Navy via author's collection)

The first production airplane that had been ordered in September 1947 was accepted in August 1949. CAPT Hayward flew it up to Moffett Field to show it to VC-5 personnel and then ferried it to Patuxent River to begin the Navy's evaluation.

While the AJ was fully carrier-capable and could be folded for stowage on the hangar deck, the folding process was time consuming. It involved the attachment of a ladder to the aft fuselage and external hinges and hydraulic actuators to the wings and vertical fin. When tested by the Navy during BIS trials, a trained team of four men, working quickly, took 16 minutes to complete the folding process while the airplane was parked on the ground, not on a moving ship, and with no wind. "Under normal shipboard conditions, the auxiliary folding mechanism must be hoisted to the wing by rope and under adverse conditions of cold weather, high winds, wet wings or rolling deck, the wing and tail folding time may easily double or triple the time reported in the above tests."[6] While meeting the letter of the original requirement and minimizing weight, it was to prove one of the least attractive features of the AJ as far as the carrier's captain and air group commander were concerned.

The production AJ-1 didn't meet all of the original specification performance requirements. With a 7,302-pound Mk 15 bomb but no drop tanks, it had a takeoff gross weight of about 47,000 pounds versus the 43,000 pounds specification. Its maximum speed was 449 mph (390 knots) at 34,000 feet versus 500 mph at 35,000 feet. On internal fuel, it had a combat radius of only 460 nm compared to the desired 600 nm. With full tip tanks, however, the combat radius was 720 miles. With the lighter 3,600-pound Mk 5 bomb, full tip tanks, and a 500-gallon bomb bay tank, it had a takeoff gross weight of 51,000 pounds and a mission radius of 1,010 nm.

The AJ's reputation initially suffered from its hasty development and overly complicated systems. Two of the three XAJs were lost as a result of in-flight structural failure in NAA flight test. In an October 1952 assessment of a VC-7 accident report, which was attributed to pilot error, it was noted that of 58 XAJs and AJ-1s, 11 had been lost to that point, most due to mechanical failure. There were almost four AJ crashes a year for its first seven years of operation, almost 70 percent of them involving fatalities. The causes were varied and not always determined, but the failure-prone hydraulic system and the jet installation were suspected to be the source of in-flight fires in some cases. The rate of loss diminished eventually as redesigns eliminated failure modes, but the sacrifices made to take the atomic bomb to sea were significant.

Training and Readiness Demonstrations

The Navy suffered a setback to its plans in mid-1949 when the new secretary of defense, Louis Johnson, canceled the authorized super carrier *United States* (CVA-58). However, it was able to maintain the momentum of the programs that it had established to put nuclear weapon-capable bombers on its existing carriers until the Korean War reestablished the need for aircraft carriers.

Following its failure to convince Congress of the necessity for a super carrier, the Navy redoubled its efforts to demonstrate its newly acquired atomic bomb delivery capability. The Navy invited the Secretary of Defense, high-ranking civilian and military leaders of the Air Force and Navy, and other civilians including the press to view Navy air power demonstrations on 26 September 1949 from *Franklin D. Roosevelt* and *Midway* off Norfolk. At the conclusion of the day, CAPT Hayward took off from *Midway* with Secretary of Defense Louis Johnson in the copilot's seat for his return to Washington.

Continuing the training and demonstrations, CDR Dick Ashworth took off from *Midway*, at sea off Norfolk, on 5 October 1949 and established a record for distance flown by an aircraft launched from a carrier. After takeoff, he and his crew flew across the Caribbean to Panama, then to Corpus Christi, Texas, and on to San Diego for landing. Total distance flown was 4,880 miles in a flight time of 25 hrs 40 min. CDR Thomas Robinson beat his record in February 1950 with a 5,060-mile mission that took 26 hours. He launched from *Franklin D. Roosevelt* off the coast of Jacksonville, Florida,

An AJ-1 wave-off and go around on 31 August provides a good look at the large flaps and nose up trim used for landing. (National Archives 80-G-418610)

and after overflying Charleston, South Carolina, turned south and flew over the Bahamas on the way down to the Panama Canal. He then flew back north, finally landing at San Francisco.

P2V and AJ operations at sea merged in April 1950 when both aircraft were aboard *Coral Sea*, the AJ on a carrier for the first time. The event also reportedly marked the first time a P2V was launched with an Mk I in its bomb bay. On 21 April 1950, LCDR R. C. Starkey was the pilot and the gross weight was 74,668 pounds, breaking Hayward's record. CAPT Hayward and his operations officer, CDR Eddie Outlaw, had already taken off in two AJ-1 Savages that had been craned aboard with the P2V.

VC-5 was relocated to Norfolk, Virginia, in mid-1950 following commission of VC-6 at Moffett Field, California, in January with CDR Dick Ashworth as commanding officer of aircraft and men transferred from VC-5. VC-7 was then commissioned at Moffett and VC-6 moved to NAS Patuxent River, Maryland, in August 1950. (Within a year of commissioning, VC-7 was also relocated to NAS Norfolk since the early mission tasking for nuclear deliveries was Eastern Europe.)

On 31 August, CAPT Hayward began AJ and VC-5 carrier qualification at sea by making two landings aboard *Coral Sea*. An indication of those more casual times, the right seat was occupied on the first by VADM Felix B. Stump, Commander Naval Air Forces Atlantic Fleet. Qualification of the rest of VC-5 pilots followed, with the limited number of available AJs forcing the acceptance of two landings or even just one for qualification rather than the usual six.

In October 1950, VC-5 flew their AJs to Guantanamo Bay, Cuba, for training with *Franklin D. Roosevelt*. The joint exercises were going well

The system to arrest the AJ was basically the same as for aircraft half its weight. If not stopped by the tailhook engaging a cross-deck pendant, the next line of defense consisted of three Davis barriers, which lift up cables to snag the main landing gear when actuated by the nose gear or the horn immediately in front of the windshield. This was one of the first at-sea landings and occurred on Coral Sea on 31 August 1950. (U.S. Navy via author's collection)

The view from the Yorktown's plane guard helicopter of an AJ-1 landing in April 1953. Note that prudence dictates a clear deck for AJ operations whenever possible. Another AJ is just turning downwind in the pattern. (National Archives 80-G-481315)

Conveniently for AJ operations, the two catapult tracks were staggered because the catapults themselves were mounted athwart ships, allowing AJs to be positioned on both catapults at the same time. Note that the AJ-1 on Lake Champlain's left catapult has its jet intake door open. (U.S. Navy via author's collection)

when, on 27 October, an AJ dove into the water right off the bow on take-off with only one survivor, the bombardier/navigator. The cause was probably the inadvertent engagement of the control lock but may have been an elevator-boost problem subsequently discovered. As a result of this and other incidents, on 21 November, the AJs were grounded. This was the start of a 12-month period during which the AJs were grounded for a total of eight months. The fleet would be modified and the grounding lifted, only to be reinstated following another crash.

Operational Deployments

Franklin D. Roosevelt sortied from Norfolk, Virginia, to the Mediterranean in January 1951 with a Special Weapons Unit and at least the Mk 4 aboard. After completion of safety-of-flight modifications resulting from the November grounding, the Savages proceeded separately. In early February, three P2V-3Cs and six AJs departed Norfolk for Port Lyautey, French Morocco,[7] via Bermuda and the Azores. AJ operations from *Franklin D. Roosevelt* commenced in late February. On 6 March, however, an AJ was lost and the crew killed following an in-flight fire in the vicinity of the ship. As a result of this incident and difficulties encountered during operations at sea, the AJs were based ashore for inspections and modifications, including conversion from standard hydraulic fluid that was flammable to water-based Hydrolube.

The P2V-3Cs were theoretically available for missions but, if carrier-launched, would have had to be craned aboard from dockside or a lighter. There were also few Mk Is, if any, and no Mk 8s at the time, which were the only weapons that the Neptunes could carry.

Mk 8s began to be available in January 1952, but by then the P2V-3Cs were being reassigned and reconfigured. Their primary use was to maintain crew proficiency during the frequent and prolonged AJ groundings and for logistic flights.

Although the AJ was smaller than the P2V when not folded, it was still large enough to affect flight-deck operations and heavy enough to strain the capability of the deck crew to move it around. As a result, it was usually in the way and not very popular. It was also maintenance intensive, the equivalent of two Corsairs and one Panther in terms of engines and far more complicated in terms of avionics and hydraulics.

The operational approach that developed in consideration of the AJ's limitations was to base the AJs ashore with the atomic bombs stored on the deployed carriers. In the event of need, the AJs would be flown to the carriers, the bombs loaded, and a strike launched. The AJs would periodically fly out to and operate from the carriers to maintain proficiency and provide a measure of normalcy to the process so an AJ's presence on a carrier would not signal the prospect of a nuclear strike.

The VC-5 AJs deployed to Port Lyautey finally resumed operations in May, flying to and operating from *Coral Sea* and *Oriskany* (CV-34). The operations to and from the *Oriskany* were the first for the AJs aboard an Essex-class carrier. However, the AJs were again grounded in June to undo the hydraulic fluid change to Hydrolube because of an epidemic of hydraulic failures. The AJs were back in the air in October in time to participate in an Atlantic Fleet exercise operating from *Midway* in November 1951 in the Caribbean. Nine strikes were flown, four during the day and five at night, all successfully

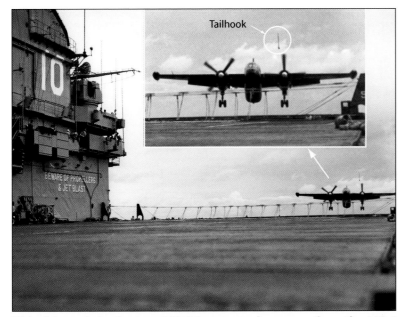

This AJ crash in August 1954 began to happen when the VC-6 pilot got too low on approach and hit the tailhook on Yorktown's ramp, breaking it off. Note that there are two Davis barriers up. What follows is one of the reasons why the big Savage was not popular with the ship's captain and air boss. Fortunately, there were no other airplanes forward as there normally would be during a deployment. (National Archives 80-G-647377)

Because the AJ contacted the Davis barriers at a significant angle, first heading to port and then starboard, the leading propeller was able to cut the cable of the first barrier and the actuating webbing of the second. With the remains of the actuator straps dangling from the nose gear, the pilot is headed to starboard, trying to straighten out. (National Archives 80-G-647385)

Unable to straighten out or stop, the AJ goes over the starboard side. Fortunately, all three crewmen were able to get out uninjured and be picked up by the plane guard helicopter within two minutes of going in the water. (NationalArchives 80-G-647394)

All three crewmen are out of the AJ and waiting for the helicopter pickup. (National Archives)

dropping weapons and returning to base without being ruled as shot down by the defending force. However, the results were graded as "Not Outstanding" for some reason.

The ASB-1 bomb director finally became available in 1952. Up until then, the AJ bombardiers, like the P2Vs, were using the APA-5, which provided poor target definition and was unreliable at high altitude. It was originally developed for an attack on a ship at sea, which was a readily discernable radar target. It had been modified for use at high altitudes against onshore targets but not satisfactorily.

A VC-6 detachment was established at NAS Atsugi, Japan, in March 1953 to provide nuclear strike capability to the 7th Fleet. Operations from the carriers were accomplished on a regular basis for crew proficiency, but the AJs never remained overnight except occasionally for a weapon-loading exercise or more often a mechanical problem that, worst case, required the carrier to steam to Atsugi to offload it. From the BIS Carrier Suitability report:

> *To observers on the deck, the size of the airplane in the approach tends to give a misleading impression because it appears to be closer and slower than the conventional carrier-based plane in similar position. Since the airplane does not respond rapidly to control movements, the Landing Signal Officer's signals should be kept to a minimum and he must decide early in the final approach whether or not the airplane is in a satisfactory approach condition and can be safely brought aboard. No late wave-offs should be attempted. In addition, corrective signals such as the "low dip" and the "high dip" should be avoided immediately before the cut. Response to these signals will result in a climb at the cut... or cause the airplane to settle at the ramp. If the airplane is slightly fast, but in a steady approach, it is better for the Landing Signal Officer to hold a "Roger" signal and give an early cut rather than give a "fast" signal in an attempt to slow the airplane just before the cut.*

> *The landing should be made with the airplane in a slightly nose high attitude so that the nosewheel is just clear of the deck. The impact load of the landing is absorbed by the main landing gear only, and the tail is kept well clear of the deck. Attempts at full flare landings will frequently result in damage from tail grounding...[8]*

AJ-2 Program

The Navy ordered 55 new and improved Savages, the AJ-2, in February 1951. The AJ-2 had a maximum gross weight of 54,000 pounds for catapulting and an arrested landing limit of 37,500 pounds. The major changes, based on AJ-1 flight test and initial operational experience, were to the empennage as described earlier, the hydraulic system, and the cockpit. The hydraulic system was simplified, but not all changes were retrofitted to the AJ-1. The AJ-2s also had a revised cockpit and canopy so the third crewman was seated on the upper deck, facing rearward so he could provide a lookout for enemy fighters, and the emergency egress hatches were enlarged.

At the behest of the patrol pilots who had been assigned to fly the multi-engine AJs, the stick was changed to a control wheel and the throttles were moved to the center console. The wheel also improved handling qualities in single engine and hydraulic-failure emergencies. The right-hand throttles made it possible for the bomber-navigator to help with the piloting duties, since this throttle location and the autopilot allowed him to control the aircraft. This control location philosophy was carried over to the A3D.

The AJ-2's first flight was 19 February 1953. These were built at the North American plant in Columbus, Ohio, rather than at Downey, California. The last AJ-2 was delivered in December 1953.[9]

An AJ-3 was proposed in November 1952. It addressed the operational issue on an Essex-class carrier of an AJ going hard down just before launch. Even when folded, the AJ did not fit on the forward centerline elevator, so the launch process had to be suspended while the dud was moved to the deck edge elevator abeam the island. The AJ-3 would have had a powered wing fold positioned 4 feet farther inboard, a folding nose, and a reduction in the length of the vertical fin/rudder with powered folding. It also incorporated a retractable "ram duct" jet engine inlet and a 6,350-pound thrust J33-A-16A for improved performance with the jet engine running. BuAer politely rejected the unsolicited proposal in a letter to North American dated 27 January 1953.

A relatively straightforward re-engining of the AJ with T34 turboprop engines was proposed to BuAer by North American in November 1953. The schedule projected the first "AJ-4" delivery in May 1955 of

A VC-6 AJ-1 is being offloaded from Yorktown onto a barge in the harbor at Sasebo, Japan, in October 1953 following a hard landing that resulted in failure of the port engine mount. Because of the time and effort it took to fold the wings and vertical tail, an AJ wasn't normally folded while aboard unless it had to be struck below, like this one, until it could be removed. (U.S. Navy via author's collection)

an AJ-2 modified during a periodic overhaul. The program was recommended by the Commander Heavy Attack Wing One in a memo dated 5 January 1954. It noted that in addition to allowing deck runs in lieu of catapult takeoffs, the additional power provided by the turboprop engines would make possible the substitution of "tail protection armament" for the jet-engine installation and allow a faster climb through icing conditions, which "would greatly offset the lack of de-icing equipment in AJ aircraft." The memo was careful to avoid criticizing the forthcoming A3D (to be described in Chapter 5) but noted that it would be restricted to carriers with the steam catapult. It also pointed out that "The turbo-prop AJ would largely retain an airframe which has become relatively dependable as a result of several years of intensive 'de-bugging', maintenance training, accumulating usage data on spare parts and stocking of supply channels. It may well require a like period to attain this degree of operational reliability with the A3D."

The Heavy Attack Wing's recommendation was endorsed by the Commander Air Force, Atlantic Fleet, in a memo dated 15 January 1954. It noted that the AJ would "retain a worth-while combat radius at low altitude" while the A3D's combat radius when flown at low altitude would be "a small fraction of its radius at altitude." It went on to state "The all weather capability of the A3D is questionable from a structural point of view because of its relatively low design load factor and the grave possibility of exceeding the airplane's strength in rough air. The AJ, with a 50 percent higher design load factor and an airframe with considerable satisfactory service experience behind it, can be considered truly all weather."

The AJ was finally more welcome on board after the advent of the angled deck and in-flight refueling. As BuAer noted in a memo

After incidents when the Davis barriers failed to stop nose gear-equipped airplanes as intended, the barricade—a system of vertical straps linked to upper and lower cross straps—was added forward of the barriers. This increased the likelihood that an un-arrested airplane would be stopped before it crashed into the pack parked forward on the carrier. This AJ-2 is being used to evaluate the interaction of the Savage and barricade. (U.S. Navy via author's collection)

If catapulted with the nose wheel swiveled backward, the AJ's nose landing gear tended to jam on retraction. This is why the nose-wheel tire was painted white on the right side as a pre-launch check that it was properly aligned. However, a jam didn't preclude continuing with the mission, as evidenced by this inflight refueling of an F9F-8P. (U.S. Navy via author's collection)

The angled deck minimized the AJ's shortcomings at the same time that the addition of inflight refueling made it welcome as part of the air group. Note that this AJ undergoing engine maintenance is parked aft of the island clear of the foul line on this Essex-class carrier, Hancock (CV-19), so it does not impede operations or have to be respotted between launches and recoveries. (U.S. Navy via author's collection)

to the Chief of Naval Operations dated 21 January 1954:

> *Current studies in this bureau indicate that air group compositions in the period 1956 to 1960 will be very difficult to handle in bad weather unless some emergency airborne refueling is made available for use by task force commanders. The jet fighters with short cycle times will pose a serious problem except where these fighters are equipped for in-flight refueling and airborne tankers are standing by. Among available aircraft, the AJ type airplane has the most satisfactory characteristics for use as a tanker. The magnitude of the problem is such that full consideration should be given to the continued use of this airplane as a tanker.*

This was certainly true in the Pacific Fleet. In response to a projection that the number of AJs available to the Pacific Fleet would drop to 16 from the desired level of 24, the Commander Air Force Pacific Fleet requested in March 1955 that BuAer come up

with "a practicable emergency means whereby the required number of operating aircraft of this extremely important type can be made available until the A3D is operating satisfactorily in fleets." In a subsequent message 10 days later, the Commander in Chief of the Pacific Fleet asked the Commander of the 7th Fleet for a recommendation "relative to the advisability and feasibility of maintaining embarked a suitable part of VC-6 det. Initially two aircraft and 3 crews on one carrier with suitable rotation from parent det is envisaged."

Once the angled deck carriers became available, the Commander of the Pacific Fleet Air Force requested that two AJs be deployed with each air group on the Essex-class carriers commencing August 1956 and that NAS Atsugi be provided with heavy maintenance capability and spare AJs.

In June 1957, the CNO authorized the retention of the AJ-2 in excess of VA(H) requirements for operation as tankers "as long as service life permits." However, "no further overhauls or interim rework are planned for this model." All were stricken as of October 1959.

CHAPTER FOUR

DISAPPOINTMENTS

Because of reliability and schedule problems with its Allison T40 engine, the A2J did not have a chance to show what it could do compared to its turbojet engine-powered competition. (U.S. Navy via author's collection)

Like some sons of great men, the second generation of attack airplanes from Douglas (A2D), North American (A2J), and Vought (A2U) were disappointments. In this case, however, all were done in by their engines, the A2D and A2J by the Allison T40 and the A2U by the Westinghouse J46. As it turned out, none of them were missed, the Navy having alternatives in development or substitutes readily available.

In two other programs, the Navy disappointed the manufacturers. An arguably successful attack capability, Vought's Regulus, was qualified for carrier operations but only made two deployments. Spurned by the carrier-based Navy, its deployments on submarines and cruisers foreshadowed Polaris and the cruise missile. Also included in this chapter is a brief summary of an alternative to carrier-based attack aircraft, the Martin P6M seaplane. After a troubled start, it was ordered into production, but budget constraints resulted in its cancellation.

Allison T40 Turboprop Engine

In the late 1940s, some in Naval aviation, particularly in the Power Plant Division within the Bureau of Aeronautics, were not convinced that the jet-propelled plane was suitable for carrier operations. The turboprop or turboshaft engine was an attractive alternative, combining the power of a jet engine with the efficient thrust of a propeller. In a turboshaft engine, a second turbine in the jet-engine

The failure of the Allison T40 turboprop engine affected manufacturers and the Navy even more than the Westinghouse J40 jet engine debacle, because there was no alternative as there had been with the J40. The T40 combined two power sections with a large gearbox. The length of the drive shafts between power sections and gearbox varied with the application. This is the engine for the P5Y seaplane. (U.S. Navy via author's collection)

exhaust turned a shaft running forward through the engine that turned a gearbox that turned a propeller. (The gearbox was necessary to reduce the high RPM of the turbine to one more suitable for a propeller.) Another advantage was that for the same weight, the turboprop produced more power than a reciprocating engine and with less mechanical complexity.

While not providing enough top speed for fighter applications due to the size of the propeller disc, the turboprop was of significant interest to the attack community. It promised an increase in speed and payload with less of a penalty on range and endurance than a pure jet would impose. The Navy also hoped that the static thrust of a big counter-rotating propeller (to eliminate torque-imposed control problems) driven by a powerful jet engine would obviate the need for catapulting attack airplanes weighing 50,000 pounds or more, since the existing catapults were inadequate for that size class.

Beginning in December 1945, BuAer funded the Allison division of General Motors to develop the T38 turboshaft engine and then the T40, which was two T38 turboshaft engines driving a single gearbox that turned counter-rotating three-bladed, 14-foot-diameter propellers. Either or both engines, called power sections, could turn the gearbox, allowing one-engine-inoperative safety and cruise efficiency. Its takeoff rating was 5,500 horsepower. In early 1947, after a competition among Allison, Pratt & Whitney, and Westinghouse, it was selected to power the Navy's big new patrol seaplane, the Consolidated Vultee P5Y.

In late 1948, the engine was projected to weigh 2,500 pounds, providing a weight-to-power ratio of 0.455. Other airplanes that were to incorporate the engine included the A2D and A2J. Two VTOL fighters were also planned around the T40, the Lockheed XFV-1 and the Consolidated Vultee XFY-1. All were to fly, but only the R3Y, a transport version of the canceled P5Y, was to be used operationally and for only a limited and troubled time.

Each T38 power section had a 17-stage compressor, eight combustors, and a four-stage turbine. The two-step gearbox reduced the engine's output RPM 15.7 to 1, resulting in a propeller speed of 868 rpm at normal engine speed. An electronic governor maintained propeller pitch for a constant RPM within the normal power range. There

Most of the Naval Ordnance Test Station's ideas were very successful and well received. One exception was this 5-inch rocket "gun" evaluated on an XBT2D in the late 1940s. It launched 5-inch spin-stabilized rockets. Nineteen rockets could be carried in each wing in addition to the existing 20mm cannon. The large opening in the lower surface of the wing vented the rocket blast. It was not adopted as a weapon in the Skyraider or any other attack airplane. (U.S. Navy via author's collection)

The Navy evaluated a Republic Aircraft design study of a swept-wing F-84 powered by the T40 turboprop engine as an attack airplane. The Air Force elected to fund it as the XF-84H. Two were built, with the first flying in July 1955. It had a single three-bladed, high-speed propeller, which may have been even louder than the counter-rotating prop on the A2D and A2J. The two prototypes only flew a dozen or so flights before the project was terminated in late 1956. (National Archives 342-B-01-034-3)

WRIGHT R-3350 ENGINE INSTALLATION IN AD AIRPLANE.

WRIGHT R-3350

GENERAL ELECTRIC TG-100 TURBINE ENGINE INSTALLATION WAS BUILT FOR FLIGHT TEST IN AD AIRPLANE BUT WAS NEVER FLOWN AS FLIGHT ENGINE WAS NOT DELIVERED.

G.E. TG-100

COMPOSITE JET AND TURBINE ENGINE ARRANGEMENT DEVELOPED BY DOUGLAS BUT NOT BUILT. THIS ARRANGEMENT UTILIZED TWO WESTINGHOUSE 24C JET ENGINES ARRANGED SO THAT THEIR EXHAUST WOULD DRIVE A TURBINE WHEEL, WHICH IN TURN DROVE TWO OPPOSITE ROTATING PROPELLERS.

COMPOSITE ENGINE

JET POWER PLANT CONSISTING OF TWO WESTINGHOUSE 24C ENGINES THAT SHOWED VERY GOOD PROMISE BUT DID NOT SATISFY TAKE-OFF AND ENDURANCE REQUIREMENTS.

TWO WESTINGHOUSE 24C JETS

THE FINAL POWER PLANT CONFIGURATION AS USED IN THE A2D CONSISTS OF A DOUBLE ALLISON T40 POWER PLANT DRIVING A COAXIAL PROPELLER THROUGH A REDUCTION GEAR.

ALLISON T40 PROP JET

Several different engine alternatives were considered to replace the Wright R-3350 in the AD Skyraider. The Allison T40 was selected for what would become the A2D Skyshark. (Author's collection)

were two throttles in the cockpit for each T40. The clutch between each power section and the T40's gearbox was controlled by the pilot. The first power section was started while declutched from the gearbox. Its clutch was engaged at 82 percent engine RPM, at which point the propellers began turning. The second engine could then be clutched to the gearbox and started.

The T40 design was stretching the state of the art. The electronic propeller RPM control system incorporated 25 vacuum tubes and was consequently unreliable, particularly given the vibration levels generated by the engine. The gearbox lacked durability. Unfortunately, there was no alternative engine being qualified as there was with the equally unsuccessful Westinghouse J40. BuAer did provide a contract in late 1950 to General Electric to develop the XT34 gearbox for a counter-rotating propeller but apparently canceled the effort.

The first flight engines were to be delivered in mid-1948. However, the engine was not even run at full power until October 1948. After several failed attempts to complete the 50-hour flight test qualification of the P5Y engine configuration, Allison was successful in September 1949. Flight engines were not delivered to Consolidated Vultee for the P5Y until December. Its first flight was finally accomplished in April 1950.

The P5Y was eventually canceled for being too big and expensive, but a transport derivative, the R3Y *Tradewind*, powered by the short shaft version of the T40, was ordered in August 1950 in recognition of the need for more and faster airlift in the Pacific to support the conflict in Korea. The first one flew in February 1954 with the engine finally being qualified for production in June. Eleven were built and used as transports and in-flight refueling tankers for a few years. However, the gearbox and propellers never achieved a satisfactory level of reliability. *Coral Sea Tradewind* (so few were built that most, like hurricanes, were given a name) suffered a runaway propeller and

was crash-landed in San Francisco Bay in May 1957. In June, one had a propeller problem shortly after departure from Hawaii for NAS Alameda, California, that required an emergency landing. The last straw was an in-flight gearbox failure on *Indian Ocean Tradewind* in January 1958 that resulted in a propeller assembly coming off the engine and tearing a hole in the hull. The crew was able to limp into Alameda and crash land in San Francisco Bay. That incident resulted in the grounding of the fleet for good.

Douglas A2D Skyshark

By early 1947, Douglas Aircraft Co. had spent almost two years doing design studies and mockups of attack airplanes powered by

Before the selection of the T40 engine for the AD Skyraider successor, a General Electric TG-100 engine was incorporated in a detailed installation mockup for a modification of the Skyraider, and then designated the XBT2D Dauntless II. (U.S. Navy via author's collection)

turboshaft engines. The attack class desk in BuAer had developed the final requirement for the new attack airplane that was approved and issued by the CNO in April 1947. In addition to the usual attack capability, it called for the ability and armament to defend itself from enemy fighters, operation from escort-class carriers, and a combat radius of 600 nm with a reduced bomb load. Douglas won the subsequent competition and received a letter of intent in June 1947 for design definition, wind-tunnel tests, mockups, and other activity prior to detail design and development. The new aircraft was initially designated AD-3, but this was soon changed to A2D in recognition of the fact that it was a major change, projected to have significantly better performance than the Wright-powered AD-3:

	AD-3	A2D-1
Gross Weight* (lb)	16,424	19,647
Engine	Wright R-3350	Allison T40
Horsepower	2,700	5,500
Combat Radius (nm)	285	420
Cruise Speed (kt)	177	299
Maximum Speed (kt)	296	433
Internal Fuel (gal)	377	500

* Full internal fuel and one 2,000-pound bomb

Production A2D-1 BuNo 125482, June 1954. (Jay Miller collection)

The 5,500 hp produced by the Allison T40 was absorbed by a pair of counter-rotating three-bladed propellers. In the production aircraft, the orange portion of the spinner would be a radome for a radar antenna. (U.S. Navy via author's collection)

The mockup review was successfully accomplished in September 1947 and resulted in a letter of intent for two XA2D-1 prototypes, with first flight to be in March 1949. The A2D bore a family resemblance to the AD in profile and was very similar in planform. One of the major differences, in recognition that the higher HP was going to result in higher speeds, was that the wing was only about 12 percent thick relative to its chord compared to the Skyraider's 17 percent. Another byproduct of the higher speed was the canopy, which was similar to Douglas' F4D design—single curvature glass was substituted for a blown Plexiglas canopy. The elevator and ailerons were hydraulically boosted. A speed brake was provided on the fuselage belly. The cockpit was air conditioned and pressurized. A Douglas-designed ejection seat was provided in consideration of the higher operating speed. The antenna for an AN/APS-19A radar was to be installed in the propeller spinner forward of the counter-rotating propellers.

There were three large bomb pylons, one on the centerline of the fuselage and one just inboard of the wing fold. The centerline station could be loaded with a 4,000-pound bomb and the wing stations, a 2,000-pound bomb or 150-gallon drop tank. There were 11 smaller pylons on *each* wing, nine on the folding section and two on the stub wing. Any could be loaded with a 5-inch rocket or a 500-pound bomb, although not all stations could be loaded with that big of a bomb at the same time. Four 20mm cannons were also carried.

Due to delays in T40 engine flight qualification, with the 50-hour test of the A2D configuration not being completed until early 1950, first flight by Douglas test pilot George Jansen wasn't accomplished until May 1950. It was very brief, truncated after only two minutes due to high levels of low-frequency vibration, a harbinger of the engine problems that were to plague flight test for the next few years. The second and third flights weren't much longer and resulted in an engine change, with no further flying between early June and mid October.

However, because of the performance and mission capability projected for the A2D, 10 production aircraft were ordered in June 1950.

Martin tried to take advantage of the A2D program delay caused by T40 qualification problems by proposing a twin-engine, two-seat, jet attack airplane with the Martin rotary bomb bay in March 1949. However, since the engines were to be almost as troublesome as Westinghouse J46s, it would have been out of the frying pan and into the fire as far as the Navy was concerned. (Glenn L. Martin Maryland Aviation Museum)

Pictured is the first XA2D Skyshark. It was lost in a fatal crash in December 1950, killing the Project Officer, LCDR Hugh Wood. (U.S. Navy via author's collection)

The onset of the Korean War resulted in additional orders being placed: 81 more airplanes in August 1950 and 250 in February 1951.

One problem with operating the T40 turboprop engine was thrust management. An unusual caution was provided in the XA2D-1 Pilot's Handbook revision dated 1 April 1953:

The choice of engine speed for a landing approach must be a compromise between minimum thrust and time required to regain maximum thrust. Experience will dictate the most desirable speed to be used. An emergency acceleration should be made by opening the power section controls very slowly at first and then more rapidly as the rpm increases. Too rapid an acceleration is likely to result in no thrust at all.

By December 1950, the XA2D had made 14 flights, all by George Jansen, and was finally considered ready for its first Navy Preliminary Evaluation (NPE). After several flights by experienced test pilots Col. Marion Carl and CDR Turner Caldwell and the Project Officer, LCDR Hugh Wood, the NPE ended tragically. Wood unknowingly experienced a failure of one power section. Because it did not have an overriding clutch, the failed engine robbed him of the power from the other. He was unable to arrest a worsening sink rate on a planned flyby for a visual check of reported smoke and hit so hard that the main landing gear collapsed and the aircraft caught fire.

Following Wood's fatal accident, the power-management system was changed to provide a more obvious indication of a power unit failure and then to provide automatic decoupling of the failed engine. However, the decoupler came with problems of its own, adding to the overall unreliability of the engines, the gearbox, and the propeller control system.

The second prototype was taxi tested and made brief hops down the length of the Hughes Aircraft runway in Culver City beginning in June 1951. It finally flew in April 1952, 16 months after LCDR Woods' crash. Engine reliability continued to plague and delay flight test. In mid-1952, Ed Heinemann met with Allison management to review the T40 development status and plan and politely, according to him, threatened them with what was to become the A4D Skyhawk.[1]

The first flight of a production Skyshark was accomplished in June 1953. Flight-testing continued, but in October, Douglas test pilot Gordon "Doc" Livingston had a gearbox failure that resulted in the loss of the propellers in flight. He was able, aided by the residual thrust of the power sections, to make an emergency landing on the big lakebed at Edwards. By then, however, only 10 A2Ds remained on contract for the purpose of conducting an evaluation of the type. The others had all been canceled after yet another Allison failure to complete the 150-hour qualification test. In August 1954, George Jansen experienced a catastrophic gearbox bull gear failure in the same airplane and was forced to eject.

Douglas management was so frustrated by the T40 engine problems and pessimistic about Allison's ability to solve them that they recommended to the Navy that the A2D be canceled.[2] In September 1954, the production contract was terminated. Two of the A2Ds that had flown were bailed to Allison for engine and propeller tests to support the ongoing T40 development and qualification effort for the R3Y.

Low Drag Bombs

One successful innovation associated with the A2D program was a new generation of bombs. A new family of bomb shapes had to be

The production A2D had a double curvature canopy, AD-5 Skyraider type wing pylons, and 11 pylons for small bombs/rockets on each wing. (Hal Andrews collection)

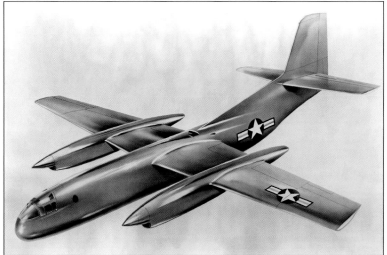

The first artist's concept of the A2J depicted a jet engine in the aft fuselage like the AJ and a cockpit located far forward. (U.S. Navy via author's collection)

developed for high-speed attack airplanes in the late 1940s. The drag of existing bombs, which had not changed significantly since before the war, was too high at the turboprop cruise speeds, even worse for jets. The box-type tail fins also began to buffet, and there were instances in which they came off the bomb in flight due to fatigue failure. Bent or missing tail fins meant that the trajectory of that bomb would be different from what was expected. In June 1947, on behalf of the Bureau of Ordnance, the Bureau of Aeronautics contracted with Douglas to develop a shape, stabilizing fins, and arming device for a new family of low-drag conventional bombs:

	Mk 81	Mk 82	Mk 83	Mk 84
Nominal Weight (lb)	250	500	1,000	2,000

The higher fineness ratio shape would also apply to external tanks and weapons pods.

Wind tunnel tests were used to develop the basic shape and stabilizing fins shown in Figure 4-1. In July 1949, Douglas used an XF3D jet fighter to flight test the new shape, using 2,000-pound bombs and 150-gallon external fuel tanks. With two bombs, the Skyknight was 51 knots faster; with two fuel tanks, 22 knots faster. Douglas estimated that the A2D Skyshark would be 53 knots faster in level flight carrying three low-drag 2,000-pound bombs than it would if it were lugging three World War II 2,000-pound bombs.

The mechanical fuses armed by rotation of a small propeller in the slipstream were replaced by a more reliable electric fuse. The new fuse greatly increased the options to the pilot for fusing for different delivery profiles just prior to the attack. The bomb was more accurate as well because the tail fins operated on a longer moment arm and were less susceptible to handling damage. Dispersion due to ballistic variation was projected to be only 3 mils for the Mk 83 versus 12 for the

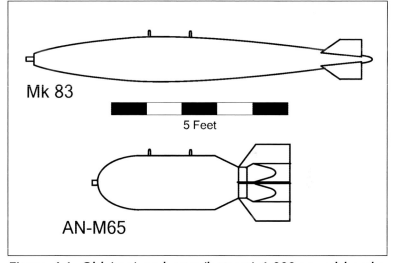

Figure 4-1. Old (top) and new (bottom) 1,000-pound bombs. (Author)

M65, for example. The new bombs were also a little lighter for the same weight of explosive.

Delivery of the new bombs began in the early 1950s.

North American A2J "Super Savage"

The XA2J program history was similar to that of the Navy's XF10F Jaguar fighter, which was being developed at the same time. Both started out as a straightforward modification of an existing design to improve performance. Both became bigger and more complicated during the design definition process. Both were terminated after a single prototype flew sporadically because its new engine was not reliable.

An elongated nose to allow for a nose turret appeared on the next iteration of the A2J along with provisions for a tail gun. The jet engine in the aft fuselage is still part of the configuration. (U.S. Navy via author's collection)

The actual XA2J was downsized with the jet engine deleted along with the provisions for the nose turret. Except for the empennage, which had to be enlarged to handle the significant increase in horsepower, the configuration was similar to the AJ's. (U.S. Navy via author's collection)

BuAer opted not to have a competition for Phase 2 of its heavy attack program. Instead, it requested a proposal from North American to do a detailed evaluation of an Allison T40-powered derivative of the AJ Savage. The sole-source approach was justified on the basis that an AJ adaptation could be done in two-thirds the time and half the cost of an all-new design. North American received a contract in April 1948 for a design definition, mockup, and wind-tunnel testing of this derivative. Almost immediately, BuAer directed North American to optimize the design for the T40 rather than minimize the changes to the AJ airframe. The first mockup review was held at North American in Los Angeles in October 1948. Shortly thereafter, North American received a letter of intent for two XA2J prototypes.

The XA2J design had grown to a takeoff gross weight approaching 72,000 pounds by the time of the second mockup review in April 1949. To reduce vulnerability to attack by enemy fighters, the design was revised to add provisions for an armed escort variant with nose and tail turrets. (The bomber was not to be armed.) To improve carrier-basing capability, complex high-lift devices—double-slotted flaps and leading-edge droop—were incorporated in the wing. The mockup review found problems with the crew ingress and egress arrangement, with the pilot provided with an ejection seat and his two crewmen, an escape slide. Even more importantly, the review board was concerned about the size and weight of the airplane. Although consistent with the requirements, the XA2J was now considered to possibly be too big and heavy to operate from the Midway-class aircraft carriers. The board was also second-guessing the decision not to have powered folding wings and vertical tail. The mockup report concluded with a recommendation that "the design

and operational requirements for this airplane be reviewed with a view toward reducing the size, weight, and complexity of the aircraft in order to achieve a design more suitable to [Midway- and/or Essex-class carrier] operation."

A few months later, North American provided a revised configuration to BuAer, which was accepted. The "special weapon" weight was reduced to 8,000 pounds, which was really a reduction in structural strength margins, since the bomb itself was not yet any lighter. Provisions for a nose turret for an armed escort version were deleted. The auxiliary jet engine was deleted. Except for the empennage, the XA2J was again similar in layout to the AJ Savage. The extra power of the 5,035-shp T40-A-6 engines (plus 1,225 pounds of thrust each) dictated a much larger and swept vertical fin and rudder and, more radically, the "flying tail" that provided pitch control from the movement of the horizontal stabilizer/elevators as a unit. The desire for defensive armament resulted in the addition of the armed escort's remotely controlled APG-25 radar-directed 20mm turret in the tail. Unlike the AJ, the wing was equipped with leading edge flaps for a lower approach speed and had integral powered folding capability, as did the vertical fin. Each of the crew was now provided with an ejection seat.

With these and other changes, the gross weight came down to 58,000 pounds, which meant the A2J could be based on the upgraded Essex-class carriers. A mockup review of the downsized   XA2J-1 was accomplished in September 1949. As a result of the review, the cockpit was changed to substitute the Douglas-type escape chute for the ejection seats. However, they were retained for the prototypes because of a fatal AJ crash during North American flight test.

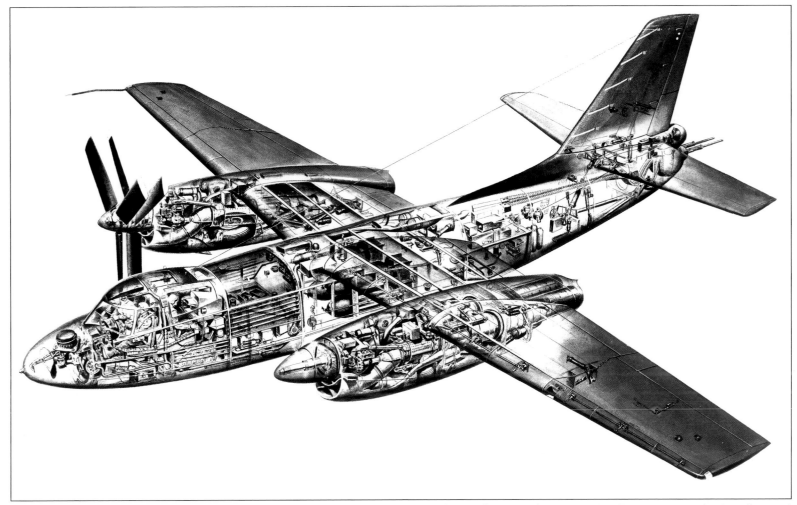

The production A2J was to have a cockpit and tail gun installation similar to the Douglas A3D's. (U. S. Navy via author's collection)

	AJ-2*	First XA2J-1**	Second XA2J-1***
Gross Weight (lb)	51,441	71,500	58,000
With Fuel (lb)	10,902	21,114	15,606
Bomb Load (lb)	7,600	10,500	8,000
Weight Empty (lb)	30,776	37,502	32,169
Maximum Speed (kt)	385	468	425
At Altitude (ft)	32,000	14,000	30,000
Service Ceiling (ft)	33,000	38,000	37,600
Combat Radius (nm)	695	1,230	1,025

* SAC dated 30 June 1957
** SAC dated 1 April 1949
*** SAC dated 1 December 1950

Engine availability delayed the XA2J program as it did the XA2D. The engines for the first airplane were to have been delivered before the end of 1949 but weren't even flight qualified until March 1951. In early 1951, North American presented an engineering study to BuAer of the substitution of the Pratt & Whitney T34 for the T40. No major changes were required other than to the nacelle forward of the wing. The weight was expected to be a little less and the takeoff performance better. The T34 also had only one power section, not two like the T40, so the engine control-system was simpler. BuAer made tentative plans to modify an early production airplane for its evaluation.

As a result of late XT40 engine delivery and problems, first flight didn't occur until 4 January 1952. However, that was still earlier than the first flight of the rival jet-powered Douglas A3D by almost a year, since it was experiencing engine delays as well. Unfortunately the XT40 engine problems plagued the flight test program as it had the A2D's. By the end of July, the XA2J had flown only five times, averaging less than one hour per flight. This rate of about one flight per month continued through mid 1953, at which point the program was terminated, in equal parts due to the progress with the A3D and the lack of progress with the T40. The impressive demonstration of the steam catapult to the

The ASM-N-4 Dove, a heat-seeking 1,000-lb bomb shown here in test on an AD-4 Skyraider, was associated with the A2J program. (U.S. Navy via author's collection)

The Royal Navy was somewhat more successful with its turbo-prop-powered attack airplane, the Westland Wyvern, powered by an Armstrong Siddeley Python engine. It did have an extended genesis, with the Python being its third engine, and was only operational for about five years. (Terry Panopalis collection)

Navy in early 1952 also eliminated any advantage that the turboprop-powered A2J had in launch performance compared to the A3D.

A little-known guided bomb associated with the A2J was the ASM-N-4 Dove. It consisted of a special nose and tail fitted to a 1,000-pound general-purpose bomb. The nose contained infrared seekers and a control system designed by the Polaroid Corporation. The tail provided steering and a wind-driven generator for electrical power for the control system. The concept was that it would be dropped and home in on a well-differentiated thermal source such as a power plant or a ship at sea. Initial testing was accomplished in March 1946 with significantly better accuracy from 20,000 feet than an unguided 1,000-pound bomb dropped at the same time. Polaroid subsequently subcontracted the project to Eastman Kodak. Twenty prototypes were contracted for in 1949 by Eastman Kodak, with testing completed by the Bureau of Ordnance in October 1952. No follow-on resulted.

The second XA2J, BuNo 124440, apparently never flew. After the test program was halted, the two airframes were stored for a year or so in a hangar assigned to North American at Edwards AFB since the airplane had potential as an air-to-surface missile carrier or aerial tanker. However, when the Air Force declared in 1954 that they wanted the hangar for other purposes, the decision was made to destroy them. The lack of progress with the T40 and the continuing potential of the A3D had eliminated what little chance of resurrection existed up until then.[3]

Vought SSM-N-8 Regulus

In addition to the piloted bombers, the Navy initiated a program in 1946 to develop an unmanned, 500-nm range, subsonic, surface-to-surface "assault missile"—a forerunner of what is now called a cruise missile. Originally intended for launch from submarines, it was to have folding wings and tail for compact stowage. Chance Vought won the competition and received a contract in late 1947. Their missile, given the name Regulus, was powered by a single Allison J33 jet engine. It was small, with no horizontal tail, but capable of carrying a 3,000-pound Mk 5 nuclear device. The wingspan was only 21 feet and the length was 35 feet. The gross weight was about 14,000 pounds. Initial development and some production aircraft had a retractable landing gear, allowing recovery and re-flight.

Initial testing was accomplished at Edwards Air Force Base, then known as Muroc Army Air Field. The controlling airplane was a TV-2 (T-33), a two-seat jet trainer powered by the same engine as the Regulus. After extensive ground testing, including high-speed taxi tests, the first flight was attempted on 22 November 1950. After take-off and during climb out, the missile went out of control, just missing the TV-2 and crashing on the lakebed. The cause was determined to be a fatigue failure of a hydraulic pump installation, probably the result of all the high-speed taxi tests that preceded first flight.

After that inauspicious start and the associated delay, flight test went fairly well following a successful flight on 29 March 1951. Out-of-sight control became standard from a ground station, with the missile being accompanied by a chase airplane capable of shooting down the Regulus if required. That could have been a problem, because at full throttle, it was faster than most of the Navy's fighters at the time. They also didn't have the unrefueled radius of action to take the Regulus to a target at its maximum range and then return.

In 1952, the CNO decided to add cruisers and aircraft carriers to the list of potential Regulus launch platforms. The cruiser Navy was enthusiastic, but the carrier Navy was not. Nevertheless, 10 carriers were eventually configured to carry and launch the missile in addition to four cruisers.

Operationally, the Regulus was to be blasted into the air off a short inclined rail launcher by two 33,000-pound thrust JATO bottles. The first launch by JATO was accomplished on 31 January 1952 from Point Mugu with missile control transferred to the submarine *Cusk* for the flight. Although reversion to control by the chase aircraft turned out to be required because the missile did not respond to the submarine's signals, the mission was concluded as planned by a supersonic dive into the target.

The Regulus initially had no internal guidance or target acquisition capability. It was to be guided to the target and detonated by radio command. Since it had a range of 500 nm, various combinations of ship and aircraft control were employed, to include handoffs to submarines and/or airplanes down range. The first operational control airplane was the F2H-2P. Removal of the cameras provided space for the Regulus control system, and it could be reconfigured aboard ship for either its photograph or controller mission. The two Banshees that were the original operational control airplanes were soon replaced in fleet operations by faster F9F-6Ds, which were in turn replaced by FJ-3Ds beginning in May 1956. Since the operational control airplanes were all single-seat, the pilot had to fly his airplane as well as provide guidance commands to the missile.

Since loss of guidance was not unknown and having an atomic bomb wandering around unsupervised was unwise, an onboard back-up system, Positive Flight Termination, was developed. It utilized a set of sequential times and checkpoint signals. In the event that control was lost, it would maintain heading and cause the missile to go terminal when the clock ran out, hopefully in the vicinity of the target. It was incorporated in the Regulus fleet in 1960.

Qualification for carrier launches followed, with the first accomplished from *Princeton* in December 1962 using the cumbersome flight-test rail launcher. In August 1954, the operational evaluation began. Four boosted launches were conducted from *Hancock* (CV-19) in October 1954 and March 1955. On the last of these, the Regulus control was transferred to a submarine. Successful launches were also accomplished from two other Regulus platforms, the submarine *Tunny* and the heavy cruiser *Los Angeles* (CA-135). The latter deployed with three nuclear-armed SSM-N-8s to the western Pacific in 1955, marking its first operational availability.

The carrier-based mission was first assigned to VC-61, a photographic-reconnaissance squadron then based at NAS Miramar, California. Now that the steam catapult was available to produce the requisite 180-knot end speed, a wheeled cart was used to catapult the Regulus like an airplane. The operational suitability testing was accomplished in August 1955 aboard *Hancock* steaming off Hawaii, just before the start of its deployment to the western Pacific with Regulus. The first shot was a success; with the missile controlled from an F9F-6D to a direct hit on Kaula Rock. At least three more missiles were fired in exercises during the deployment, with the Regulus being used as an air-to-air target for the VF-121's F9F-8s after its simulated attack on a target.

The major drawback to the Regulus aboard the carrier was that it was a one-time weapon like the TDR and F6F drones. Even though it was small and had folding wings, when the checkout vans, launchers, and chase airplanes were taken into account, it took up the space of a squadron of airplanes. Piloted airplanes could fly multiple missions, and in the unlikely event that the dreaded nuclear card had to be played, might well be as likely as the Regulus to reach the target and potentially more accurate.

Operational testing was accomplished from Hancock *using a mobile rail launcher and JATO booster. Pictured is one of the first Regulus I launches in October 1954.* (Vought Heritage Ctr.)

In September 1955, Guided Missile Groups One and Two were established on the West Coast and East Coast, respectively, to provide an en route and/or terminal guidance option for missiles launched from the carriers, cruisers, and submarines. GMGRU-1's first deployment aboard a carrier was a detachment operating two F9F-6Ds from *Lexington* beginning in May 1956. However, the carrier itself embarked no missiles, and no cruiser or submarine launches were accomplished during the seven-month deployment as the missiles were only to be used in a shooting war.

Five more GMGRU-1 detachments, now flying FJ-3Ds, deployed on carriers without Regulus aboard. Their mission was to provide chase and guidance, if required, for submarine-launched missiles. In fact, when the carrier first arrived in the vicinity of Okinawa, the FJ-3Ds were usually flown off and based at Naha until the deployment ended. The tradeoff of a war load of cruise missiles versus a squadron of airplanes was unappealing to the carrier Navy. The last deployment was aboard *Lexington* and ended in December 1958.

A GMGRU-2 detachment also made one deployment aboard a carrier along with its missiles when a detachment departed the East Coast on *Randolph* (CVA-15) in July 1956 to the Mediterranean. It was provided with five missiles and an Induced Pitch Launcher, which eliminated the space requirement of catapult launch carts but reintroduced the potential damage from JATO blast and the annoyance of the noxious exhaust products. However, the unit still was considered to take up too much space relative to the likelihood of a doomsday mission. No launches were accomplished during the deployment itself. It was the second and last deployment of an aircraft carrier with Regulus aboard. GMGRU-2 did not make another deployment either.

Although spurned by the carrier-based Navy, the Regulus was embraced by the subsurface and surface Navy. It provided them with

A wheeled cart was developed for Regulus I launches from carriers equipped with steam catapults. This minimized the disruption to normal carrier operations, not to mention the noxious byproducts of the JATO launch. (Vought Heritage Center)

both a new *raison d'être* and the means to accomplish it. *Los Angeles* was the lead cruiser for development of the ship installation and procedures. She made several deployments to the western Pacific between 1955 and 1961. *Helena* (CA-75), *Macon* (CA-132), and *Toledo* (CA-133) also made deployments and launched missiles for training and proficiency many times. Beginning with *Tunny's* first patrol in late 1959, five different Regulus-carrying submarines made 40 two- to three-month patrols in the Pacific. An average of two was at sea at any given time. In 1961, they became part of the first Single Integrated Operational Plan (SIOP), which assigned targets for all U.S. Navy and USAF nuclear weapons. The last patrol ended in mid-1964. After that, the remaining missiles were employed as a non-recoverable target drone, one of the most realistic available. The last flight was launched from Barking Sands, Hawaii, in June 1966.

Vought A2U Attack Cutlass

The F7U Cutlass program began in 1946 as a high-performance, twin-afterburning jet-engine fighter. It was a very unusual configuration, not having the horizontal tail that was believed to be a source of transonic handling-quality problems. The F7U-1 suffered from weight growth and engine problems. Only a handful was built. The basic concept had potential, however, so the Navy contracted with Vought for the F7U-3 with more powerful engines, albeit from the same manufacturer. It did no better than the F7U-1 from a weight and engine performance standpoint but the Navy persevered, since it had a requirement for a high-performance general-purpose fighter and both of the other two possibilities, the McDonnell F3H Demon and the Grumman F10F Jaguar, were in at least as much trouble as the F7U-3 from a development standpoint.

This fanciful artist's concept of the A2U depicts it with seven, four-shot Zuni pods attacking what appears to be a fleet oiler (used for replenishment of combat ships) armed with antiaircraft guns. (National Archives)

Because Westinghouse had not been able to increase the J46's thrust and reduce its fuel consumption as expected, the F7U-3 was short on range. Nevertheless, the Cutlass was well regarded for its air-to-ground capability and was among the first to drop a bomb while supersonic. Vought was promoting an attack variant, the V-389, as early as 1951 and trying to get the Marines interested in a shore-based version as well. Vought provided a brochure-type proposal to the Navy in 1952 that provided a shopping list of improvements to the F7U-3 as an attack aircraft. In order to carry additional stores, it suggested adding a small pylon on the inboard wing just outboard of the existing high-capacity pylon and three small pylons on each outboard wing panel. The "A2U-1" was powered by two J46-WE-8 engines. The inboard wing was extended aft along with the speed brakes to provide more wing area. The "A2U-2" had a more significant change to the wing—the fins were

moved inboard by about six inches and the outer wing panel had less chord and big "delta" ailevators that together increased the aspect ratio of the wing—and was to be powered by two J46-WE-2s. The "A2U-3" was the "-2" reengined with either two afterburning Wright J65s, a license-built derivative of the Bristol-Siddeley Sapphire, or two J47-GE-2s instead of the perennially troubled Westinghouse J46s.[4]

The F7U-3 already had an attack capability. Early on, the F7U-3 had been configured to carry a belly pod that could be loaded with 32 2.75-inch folding fin rockets. As a general-purpose fighter, the F7U-3 was also qualified to deliver conventional and nuclear bombs and had made one of the first supersonic bomb drops. Although the –3 only had two stores pylons in addition to the belly pod, two additional pylons had been added to the air-to-air Sparrow missile-carrying F7U–3M so it could carry four missiles. The F7U-3M outer wing

Rocket pods were developed so more rockets could be carried on a single weapons pylon. Here an AD Skyraider is used in March 1955 to illustrate a maximum load, with the large Aero 7D pods containing 19 2.75-inch folding fin rockets interspersed with the smaller Aero 6A pods containing seven rockets. The 14 wing pylons could therefore carry 194 of these small rockets instead of only 12 of the HVARs plus two Tiny Tims or bombs. (U.S. Navy via author's collection)

Larger, aerodynamically counterbalanced ailevators were considered for the A2U and evaluated in late 1953. These included extending the ailevators forward to the wing leading edge, possibly to provide for lower control system loads (hinge moments) and simplification of the flight-control system. The so-called delta-tip ailevators also increased the wing area and aspect ratio slightly, the former reducing the wing loading and the latter increasing the cruise efficiency. (Vought Heritage Center)

panels were also designed to incorporate additional fuel cells, since the Sparrow missile radar precluded the installation of the existing inflight refueling probe.

In early 1953, the Navy decided to authorize Vought to develop an attack version of the Cutlass, probably due to the difficulties being encountered at the time by the Douglas A2D. The Navy was in an awkward position, urgently needing to provide a backup for an essential mission requirement that could only be fulfilled by a big airplane. The A2U-1 detail specification was signed in July 1953.

The configuration was very similar to the "A2U-1" proposal except for the stores pylon changes. Instead of several small pylons on the outer wings, there was only one like the F7U-3M. There were three small pylons for small stores, but these were on the belly as an option to the rocket pod. The wing area and size of the speed brakes were slightly increased by extending the trailing edge of the wing inboard of the fins aft.

An example strike mission load was seven pods each loaded with 19 2.75-inch rockets. Maximum bomb load with full internal fuel was to be 7,000 pounds. Each of the two inboard wing pylons could carry a 3,500-pound store, while the outboard wing and outboard belly pylons were restricted to 500-pound stores. The center belly pylon could carry a 2,000-pound bomb. The avionics suite, bomb director, and gunsight were optimized for the attack mission.

One change that seemed counterintuitive for an attack aircraft was that the number of 20mm cannons was reduced from two on each

The Vought-designed refueling system consisted of two pivoting booms, which positioned the drogue somewhat farther below the tanker than the hose system did. Note that the receiver is armed with two Mk 7 nuclear weapons and the tanker with one as well, which was probably an error on the part of the artist. (Vought Aircraft Heritage Foundation)

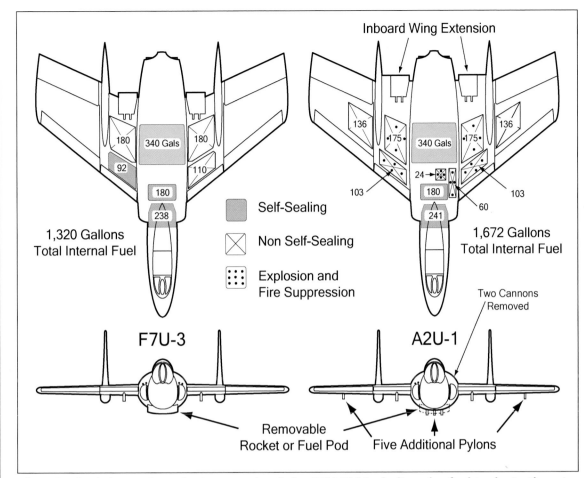

Inboard Wing Extension

136   •175•   340 Gals   •175•   136

24→ ▨

180

103   60   103

F7U-3

1,320 Gallons
Total Internal Fuel

340 Gals

180   180

92

110

180

238

■ Self-Sealing

⊠ Non Self-Sealing

⊡ Explosion and
Fire Suppression

A2U-1

1,672 Gallons
Total Internal Fuel

241

Two Cannons
Removed

Removable
Rocket or Fuel Pod

Five Additional Pylons

The A2U had the outboard wing panels of the F7U-3M including the fuel tanks in the wing and the stores pylon under the wing. Other differences from the F7U-3 included the aft extension of the inboard section of the wing and the fuel and armament systems. (Author)

1.22 originally expected from the J46-WE-2. Vought and the Navy were also hoping that further improvements to the J46 engine were forthcoming, so-called Block III. In June 1954, the A2U's radius of action on internal fuel with the Block III engine was projected to be 450 nm, twice that of the F7U-3 powered by the J46-WE-8.

Changes were made to minimize the empty weight increase. For example, four hydraulic pumps instead of six were now considered sufficient, and the original GE G-4 artificial stabilization system was replaced by a lighter Vought-developed unit. With a pilot, full internal fuel, and guns and ammunition but no stores, the gross weight was a little over 30,000 pounds. In anticipation of its use by the Marine Corps from advanced shore bases and an increase in *design* landing weight from 21,000 to 25,000 pounds (the approved landing weight for the F7U-3 was approximately 23,000 pounds), improved brakes and bigger tires were to be provided. The slightly different stance with the modified landing gear and bigger tires required a longer arresting hook.

Four development aircraft were planned, BuNos 138368 through

side to two on the right side only. However, this provided the volume for 80 additional gallons of fuel in the left-hand gun and ammunition compartments; off-setting the reduction in volume in some of the existing fuel tanks caused the addition of a fire and explosion suppression system to reduce vulnerability.

Except for the addition of the outboard wing panel tanks of the F7U-3M and explosion proofing, the basic configuration of the original fuel system was simplified to reduce complexity and weight. Including the left gun/ammo bay fuel, another 352 gallons of fuel was added over and above that in the F7U-3 for a total of 1,672 gallons of internal fuel. Like the baseline F7U-3, external fuel could be carried in a 220-gallon jettisonable belly pod and/or a 150-gallon drop tank on each inboard wing pylon.

The A2U was to be powered by two Westinghouse J46-WE-18 engines. In recognition of Westinghouse's lack of credibility with respect to delivery schedules, the engine installation was designed to also accommodate the -8 engine. The –18 was to provide the 6,100 pounds of thrust and cruising Specific Fuel Consumption (SFC) of

138371, with the first one to be completed in October 1954 for a first flight in January/February 1955. In the meantime, two F7U-3s were assigned to conduct testing of the Vought stabilization system in early 1954 and fabrication of structure for the first four airplanes was in work. Vought was planning on producing 92 A2Us as a follow-on to F7U-3M production, with the first delivery in October 1955 and the last in November 1956.

In late 1954, however, the Navy finally decided that the Attack Cutlass was not essential to its needs. On 18 November, with the first airframe complete and ready for checkout, it canceled both development and production. In its announcement, the Navy justified the decision on the J46 schedule status. Other reasons likely were the inevitable budget reductions after the end of the Korean War and progress in development of the F3H-2, which was also a big fighter with general-purpose mission capability.

In any event, most of the Cutlass deployments between 1955 and 1957 were with attack squadrons, and half of those went out with F7U-3Ms, which had the two additional wing stores stations. In effect, the Navy was deploying and employing the F7U as the A2U.

A June 1954 analysis of the benefit of inflight refueling was accomplished with an A2U bomber/buddy tanker team. Both aircraft were equipped with a fuel pod and an external 150-gallon tank. The tanker carried a refueling store with an additional 150 gallons of fuel and the bomber, a 2,000-lb nuclear store. The analysis projected an increase in radius of action for the bomber of 220 miles, or about 50 percent, with the en route refueling beginning at a distance of 175 miles from the carrier. (Author)

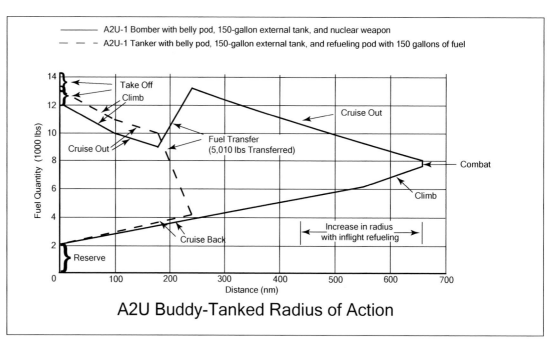

A2U Buddy-Tanked Radius of Action

P6M Water Taxi 31 January 1958. (Glenn L. Martin Maryland Aviation Museum)

Martin P6M SeaMaster

The Navy had long operated big, long-range seaplanes to patrol the oceans, scouting for enemy fleets and searching for enemy submarines. One of the first crossings of the Atlantic by an aircraft was accomplished by a U.S. Navy crew flying the Navy-Curtiss (NC) TA, for Transatlantic Attempt. Four NCs were built for the attempt, three began it, and one, NC-4, finally reached Lisbon 19 days later.

The attraction of a seaplane was that it could be much bigger and have a longer range than any airplane that could be launched from an aircraft carrier. In 1949, the Bureau of Aeronautics created the Seaplane Striking Force (SSF) concept. The bomber component was to be a big, jet-powered seaplane that could carry a 30,000-lb payload out 1,500 miles and return. It issued a request to industry in April 1951. Both of its seaplane contractors, Convair and Martin, responded with Martin winning the competition.

Other elements of the Seaplane Striking Force were the F2Y Seadart fighter, the P5Y patrol bomber, and the R3Y Tradewind transport. The F2Y proved to have poor takeoff and landing characteristics on its hydro-skis and was eventually canceled. Two XP5Ys were built. One crashed, surprisingly not due to an engine failure, and the other never flew. As previously mentioned, a handful of R3Ys were produced and placed into service for a short time.

BuAer awarded a contract to Martin in October 1952 for two prototypes of the XP6M-1 SeaMaster. The swept-wing beauty was to be powered by four Curtiss-Wright J67 jet engines and have a top speed of Mach 0.9 at sea level. (The engine was subsequently changed to the Allison J71 with afterburner due to J67 development delays.) It was a big aircraft, with a gross weight of 160,000 pounds and a wingspan of 103 feet, approximately the same size as the Air Force's B-47 Stratojet.

First flight of the XP6M was accomplished in July 1955. All went well until 7 December, when the prototype suffered a loss of longitudinal control due to an unknown malfunction, causing it to disintegrate in flight, killing the four-man crew. Test flying with the second prototype resumed in May 1956, but it crashed in November due to a miscalculation of elevator system control loads versus hydraulic actuator capability, with the crew ejecting safely. Nevertheless, the program had enough support within OpNav and BuAer to continue, albeit from a dwindling number of seaplane enthusiasts who saw it as accomplishing missions that were being ignored, neglected, or dealt with by inferior means. Some 1954 and 1955 studies concluded that the SSF was both less expensive and more effective than the carrier task force in long-range strategic bombing missions.[5]

The first of six YP6M-1s flew in January 1958. The first production P6M-2, now powered by non-afterburning Pratt & Whitney J75s and capable of being overloaded to 195,000 pounds, flew a year later in February 1959. Unfortunately OpNav, faced with budget constraints, decided that the SeaMaster's nuclear weapon delivery capability was redundant, while that of the forthcoming Polaris missile and carrier-based airplanes could do the other missions envisaged for it, like air-to-air refueling and mine-laying. BuAer formally terminated the program in August 1959 and had the 14 YP6M-1s and P6M-2s scrapped.[6]

The XP6M-1 SeaMaster was rolled out in January 1955, supported by four simple beaching-gear bogeys, each of which was separately removed for flight once the seaplane was in the water and reinstalled for taxi out of the water. The operational version of the beaching gear was a large, variable-buoyancy wheeled rig that the P6M water-taxied onto and off of. (Glenn L. Martin Maryland Aviation Museum)

Like the A2J and the A2D, the P5Y program was delayed by T40 engine qualification problems. Here the first XP5Y sits on the beach awaiting a ship set of engines. The second prototype was completed except for engines, which were never installed. (U.S. Navy via author's collection)

The U.S. Air Force B-47 first flew in December 1947. The P6M was approximately the same size and initiated by the Navy as the first production B-47s were being delivered. Although the P6M's stated mission was mine-laying to keep Russian submarines bottled up, it was clearly capable of the strategic delivery of atomic weapons similar to the B-47's potential. (National Archives 342-B-01-001-8)

The P6M design was optimized for water takeoffs and landings. The T-tail kept the horizontal stabilizer out of most of the seaplane's wake as it accelerated for takeoff. The engine inlets were located above the wing to minimize sea water ingestion during takeoffs and landings. The wing was provided with leading edge slats to minimize takeoff and landing speed. The wingtip-mounted floats were effective on the water and yet did not produce so much drag that they needed to be retracted. (Glenn L. Martin Maryland Aviation Museum)

The P6M had a large rotary bomb bay located under the wing center section. The red outline is the water rudder for directional and speed control during water taxi. (Glenn L. Martin Maryland Aviation Museum)

CHAPTER FIVE

THE WHALE'S TALE

The A3D Skywarrior was the Navy's main battery for 10 years—happily never having to be called upon to fulfill its prime objective of atomic or thermonuclear bomb delivery—before a pair of smaller bombers supplanted it. It continued to serve in supporting roles for another 20 years. (Robert L. Lawson collection)

Even if the Allison T40 turboprop engine hadn't done in the A2J, it's all but certain that the Douglas A3D Skywarrior's jet-powered performance would have when the steam catapult and angled deck were introduced in the early 1950s. These carrier improvements, so critical to carrier basing of high-performance airplanes, were not known when the Navy committed to the program that resulted in the A3D. The big jet, fondly referred to as the Whale, probably wouldn't have been able to safely operate from the Essex-class carriers that had conventional hydraulic catapults and an axial deck. As it turned out, however, the A3D was by far the biggest and heaviest jet to operate from the 27 Charlie carriers and naval aviators did it without auto-throttle and direct lift control. At night...

Specification, Evaluation, and Selection

A jet bomber was Phase 3 of BuAer's December 1945 plan to seize a portion of the strategic nuclear strike responsibility for the Navy. At the time, its use was dependent on a new class of super carrier since it

The USS United States *(CVA-58) was intended to be the Navy's launching pad for large strategic bombers. What looks like an angled deck is actually for additional catapults aft, to shoot aircraft off left, right, and straight ahead in order to launch a strike or defensive fighters as quickly as possible. One of the aircraft types appears to be a shrunken P2V; the fighters resemble the FH Phantom and are two different sizes.* (National Archives 80-G-706108)

was projected to weigh 100,000 pounds. In 1948, the requirement was to carry a 10,000-pound bomb to targets 1,700 to 2,000 nm distant and return to its carrier. Maximum speed and altitude were Mach 0.9 and 40,000 feet, respectively.

The big bomber was originally expected to operate from the Navy's new super carrier, *United States.* With a flight deck 1,028 feet long and 180 feet across at its widest point, it was big enough to handle an aircraft complement of eighteen 100,000-pound bombers. The hydraulic catapults were capable of launching the bombers with an end speed of 105 knots. The arresting gear could recover that airplane

at an engaging speed of 125 knots. The keel was laid for *United States* on 18 April 1949. It was canceled only four days later.

Bids were requested in August 1948, with the proposals submitted in December. The Convair, Douglas Santa Monica, Fairchild, Lockheed, Martin, and Republic designs were easily dropped from consideration and demonstrated how difficult the requirements were perceived to be.[1] Convair's design resembled a Boeing B-47 with one J40 under each wing and one in the tail; like the Douglas Santa Monica three-engine proposal, it was rejected on the basis that its gross weight was already projected to be 100,000 pounds, "which leaves no margin

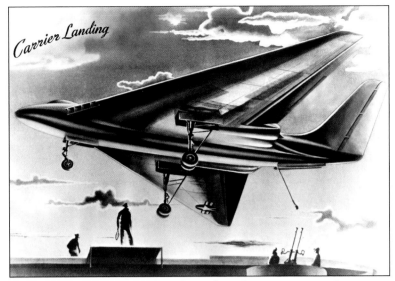

In response to the U.S. Navy's need in the late 1940s for a carrier-based, long-range jet bomber, the Glenn L. Martin Company, the successful purveyor of large seaplanes, proposed the Model 245, a hybrid that could fly to the carrier's operating area, refueling from submarines along the way, and then jettison the boat hull for operation from a carrier. It was apparently to be used only from staging from a carrier, because it did not have folding wings. (Glenn L. Martin Maryland Aviation Museum)

for correction of defects or service growth." Fairchild proposed a canard configuration powered by six J46 engines with a jettisonable biplane wing for takeoff; besides being the most unconventional and heaviest design, it was the most expensive. Lockheed didn't provide a cost quotation and it is surprising that they bothered to submit their design. It was powered by four J40s and even at a takeoff weight of 100,000 pounds had only half the mission radius desired. Martin

suggested that the Navy buy a non-compliant design. It only had a two-man crew and didn't have access to the bomb bay, protected fuel tankage, folding wings, or a turret, presumably to keep the weight below 100,000 pounds. Republic didn't even bother trying to get down to the weight limit with its proposal powered by two J40s and two J46s. It also set the record for number of fuel cells: 10 in the fuselage and 32 in the wings.

The original Douglas El Sugundo proposal was similar to the XA3D but the cockpit was farther aft and a tail-wheel landing gear was used to avoid the need to fold the tail. During the redesign following cancellation of the carrier United States, the tricycle landing gear and folding tail were incorporated. (Jay Miller Collection)

This twin J40-powered Curtiss proposal was also a finalist in the Navy's long-range jet-bomber competition. Its proposed gross weight was actually 500 pounds lighter than the proposal from Douglas El Segundo but it had less range. Ed Heinemann of Douglas then refined his design so that it was the clear winner. (*Tailhook* Magazine)

The Curtiss and Douglas El Segundo proposals merited additional consideration so both companies were awarded study contracts in early 1949. The Curtiss design was similar in some features to the B-47. It had thin, very high aspect-ratio wings with a bicycle-type landing gear. The two J40 engines, however, were located at the juncture of the wing and fuselage like those on the Douglas F3D. Like the B-47, lateral support on the ground was provided by outrigger landing gear. With a 40-ft tread, these would have limited its operation to the new super carrier, *United States*, which provided more distance between the catapult track and the deck edge than the Midway class. Another deficiency was the lack of powered wing folding; a system similar to the unsatisfactory one on the AJ was to be provided. The pilot was seated on the centerline of the fuselage with the other two crewmembers seated side by side behind him. All were provided with ejection seats but the evaluator noted that Curtiss "failed to show a method of protecting each crew member from blast and interference of others when ejected."

The performance projections favored the Douglas proposal, assuming that it could meet the proposed weight, which was well below all but Curtiss' proposal and BuAer's estimate. The evaluation was cautiously optimistic: "It is considered that the Douglas estimate, while lower than the empirical average, is capable of realization with careful detail design and rigid weight control practices." The Curtiss design was considered to be slightly faster, eight knots at sea level and 12 knots at 40,000 feet, but the Douglas had a lower stalling speed and shorter takeoff distance. The Curtiss had less range, 1,535-nm radius, compared to the Douglas' 1,650 nm, (the requirement was 1,500 nm) but was considered to have a more conservative weight estimate, "sufficient to offset the radius advantages if fuel loads were corrected."

Douglas was therefore selected in mid-1949 to proceed to build its aircraft even though its bid was somewhat higher than Curtiss-Wright's:

> The heavy attack airplane is considered of vital importance to the future of naval aviation, and compromise in an engineering choice because of cost of the experimental airplane is certainly not to the best interests of the government. The engineering advantages (of) ... makes the Douglas, El Segundo design an obvious choice from an engineering point of view. The current record of Douglas, El Segundo in designing and production naval carrier aircraft is far superior to all other competitors in this competition and is considered to outweigh a temporary cost advantage. [2]

The initial contract was for the usual two prototypes and static test article. The requirements had to be changed to reflect the cancellation of the *United States* in April and downsize the airplane so it could operate from Midway-class carriers. However, the speed and altitude performance were not to be reduced: "A reduction of combat radius may be allowed if necessary."[3] This turned out to be the case, with the A3D-1 eventually having a combat radius of about 1,200 nm with a 2,000-pound store; still a tremendous leap forward for a carrier-based airplane.

The Douglas A3D mockup featured a slightly different aft fuselage shape for fairing in the tail gun installation and a shock-absorbing tail bumper. Note the access steps at the main landing gear. These were deleted from the design, with access to the upper fuselage provided through the cockpit. (U.S. Navy via author's collection)

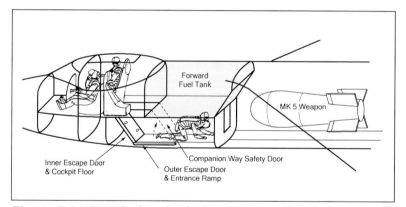

Figure 5-1. Atomic bomb designs eventually eliminated the requirement for inflight insertion of critical components. The bomb was still armed after takeoff, but manual mechanical assembly was not required. (Author)

Configuration Definition

The mockup review at Douglas's El Segundo, California, plant was held in mid September 1949, concurrent with the third review of the XA2J at North American. Douglas Chief Engineer Ed Heinemann had convinced BuAer that the A3D could meet its revised requirements at a gross weight of 70,000 pounds. The Skywarrior wing was thin and swept for speed and had a relatively high aspect ratio for range. It was mounted high on the fuselage to allow for the large bomb bay required by the first atomic bombs. The engines were housed in pods slung beneath the wing for accessibility for maintenance. Because of the height of the wing, all of the landing gears retracted into the fuselage. The crew of three was housed in the cockpit, with access provided to the bomb bay for arming the bomb in flight as shown in Figure 5-1. The access was cleverly integrated with the Douglas entry/escape chute that substituted for ejection seats as part of the all-important weight minimization.[4] Defensive armament was a remotely controlled, radar-directed twin-20mm turret in the tail. The wings and the vertical tail folded.

The Skywarrior was one of the few Navy jets that did not have provisions for drop tanks but still had excellent range.[5] The combination of a fuselage-deep bomb bay that needed to be located near the center of gravity and the volume of fuel needed by a jet airplane resulted in large fuel tanks located in front of and behind the bomb bay in addition to a tank in each wing. The engines drew fuel from the aft fuselage tank. The fuel transfer from the forward fuel tank to the aft one was automatically managed in order to keep the airplane's cg within the allowable range. The wing tank fuel was normally transferred to the forward fuselage tank to keep it full until the wing tanks were empty—they were not self-sealing so it was desirable for them to be empty and purged when going into combat. (If there were a problem with the forward tank or the transfer system, the wing fuel could be transferred directly to the aft tank.) One early A3D-1 loss resulted from a failure of the fuel balance system, with the crew having to bail out when the last 3,000 pounds of fuel would not transfer.

The crew responsibilities changed over time except for that of the pilot. In the beginning, the right-seat crewman was the bombardier/assistant pilot. He was originally likely to be a junior Naval aviator, but like the AJ, there was only a single set of flight controls, although the throttles and autopilot were accessible to the right seat. He operated the ASB-1 Bomb Director, a combined airborne search and navigation radar system and a bomb director. A periscope was included in the system as well for navigation, target acquisition, and visual bombing. A third crewman, initially described as the navigator/gunner, sat facing aft, back to back with the pilot. As with the World War II bombers, he was an enlisted man and plane captain, meaning he was responsible for the care and maintenance of the airplane. He also armed the nuclear weapon in the bomb bay after takeoff if manual arming was required. A celestial sight was located over his position and he was trained to take position sights. After a few years, the right-seat crewman was usually not a pilot and took over the navigation responsibility.

Navigation to the target was accomplished by dead reckoning—heading (corrected for estimated wind) and time, updated with visual, celestial, and radar position fixes. The attack was expected to be made with a bomb drop from high altitude using a radar-identified point.

One unusual feature of the A3D was that the electricity generators and hydraulic pumps, instead of being powered directly by a gearbox on the engine, were driven by fuselage-mounted turbines that were turned by bleed air from the engines. While this approach was qualified and operational on the Skywarriors throughout their service life, the Navy eventually decided that it and the lack of ejection seats were not to be repeated in future designs.

In its planning in the late 1940s, the Navy expected that the

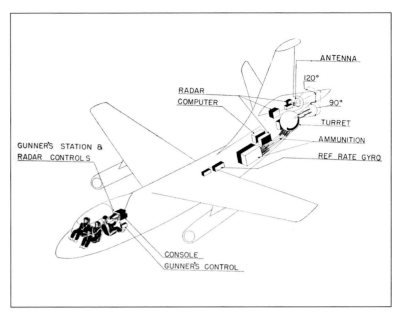

The Aero 21B Tail Turret system was a radar-controlled, automatic gun-laying system operated by the rearward-facing crewman. The turret contained two 20mm cannons with 500 rpg. The radar above the turret scanned for targets in a 120-degree cone to the rear of the airplane and could detect and lock on targets at ranges as great as 5,500 yards. The gunner could select a return on his scope for the system to track and aim the guns, which fired when the gunner pulled his trigger. Anecdotally, the system was unreliable, and realistic training was difficult to provide. (U.S. Navy via author's collection)

downscaled A2J would be in service a few years earlier than the A3D and accomplish long-range strike missions from carriers smaller than the Midway class. However, in late 1950, well before first flight of either the A2J or the A3D, the Navy had decided that it needed to expedite an AJ replacement into service. Since available funds limited the early production order to only one of the two candidates and the A3D was projected to be considerably faster and have slightly more range, it was chosen:

	A2J	A3D
Gross Weight (lb)	58,000	70,000
Maximum Speed (knots)	425	564
Mission Radius (nm)	1,025	1,075

Nevertheless, North American was to continue development of the A2J, since it would be less penalized in mission capability when operated from Essex-class carriers.

In February 1951, more than 20 months prior to first flight, Douglas had received a letter of intent for 12 production A3D-1s with J40 engines. However, the decision was made in April 1952 to change to the Pratt & Whitney J57-P-1 rated at 9,500-pound thrust. Westinghouse

The first XA3D on approach to Edwards Air Force Base in October 1952, speed brakes extended. The second XA3D had a slightly different engine nacelle, with the inlet perpendicular, not raked rearward as shown here. (Jay Miller collection)

A retractable picket fence had to be added ahead of the bomb bay due to problems with airflow within the bomb bay causing the bombs to not drop cleanly. Just ahead of the deflector is the crew access and escape door. The fairing under the fuselage adjacent to the nose wheel well covers the bombardier's periscope. The open doors under the extreme aft fuselage are for the drag parachute for field landings. (Jay Miller collection)

wasn't meeting schedule or performance commitments, putting the A3D and Navy fighter programs at risk. Heinemann convinced the Navy to switch both his F4D and the A3D to the Pratt & Whitney J57 engine for production. As it turned out, none of the airplanes that were to be powered by the J40 would become operational with it.

Development

First flight of the XA3D was accomplished on 28 October 1952 at Edwards AFB, powered by J40s. It wasn't quite ready for prime time, with the next flight being more than a month later on 4 December. It didn't fly again until January. At that point, engine problems grounded all J40-powered airplanes until July 1953. One indication of the J40 impact is that the second XA3D didn't fly until October 1953, a month after the first production A3D-1 flight powered by the J57.

The inherent lack of torsional stiffness possible in thin, high-aspect-ratio wing risked aeroelastic problems as predicted by wind-tunnel test and the newly available analog computer analysis. Fortuitously, the substitution of the J57 for the J40 and some structural changes combined to move the onset of the wing torsional instability out of the airplane's flight envelope.[6]

One of the unanticipated problems was airflow recirculation in the bomb bay and airframe buffet when the bomb bay doors were open at speeds about 200 knots. The airflow was such that bombs

Jet-assisted (actually rocket-assisted) takeoff was a requirement so that the A3D could be launched even if the carrier's catapults were inoperative. (Jay Miller collection)

Due to brake problems, a drag parachute was added for field landings. This A3D-1 is landing after setting a coast-to-coast record in March 1957. (National Archives 80-G-101774)

would occasionally not drop out of the bomb bay when released. The solution, perfected after many flights, was a fence that extended in front of the bomb bay when the doors were opened. This allowed clean drops up to Mach 0.9 at altitude.

A lift spoiler was added to the inboard wing for roll control after it was determined that at high speeds, the existing ailerons functioned as a trim tab on the torsionally flexible wing, causing roll opposite of what was commanded. (A similar problem developed with the horizontal tail.) The spoilers were retrofitted on all but 10 of the first 12 production A3D-1s and were standard on the A3D-2.

Main landing tire damage and failure were encountered early in flight test because the relatively small, high-pressure tires were prone to skid. This was addressed by the development of an anti-skid brake system that detected a skid developing and released the brake pressure momentarily to keep the wheel turning. However, landing rollout distances were still too long, resulting in a drag chute being added to later A3Ds.

Another feature was the provision for the use of Jet Assisted Take Off (JATO). Mounting points for six 4,500-pound thrust bottles were provided on each side of the fuselage just ahead of the speed brakes. In the event the ship's catapults were not available, this would theoretically allow the deck launch of a loaded A3D, since the total thrust of the JATOs and the jet engines approximately equaled the gross weight.

The fifth production airplane was flown non-stop from Edwards AFB, California, to NAS Norfolk, Virginia, on 21 August 1954 in 4 hours 40 minutes, covering the 2,385 miles at an average speed of 510 mph. The Skywarrior cruised at 40,000 feet and a Mach number of 0.82 to 0.85. At Norfolk, it was converted to be the first of five A3D-1Qs modified from the A3D-1 production run.

The eighth production airplane was assigned to the structural demonstration program. It was dived to slightly supersonic speeds at least twice in the expansion of the flight envelope. In yet another test dive in October 1955, the horizontal stabilizer attach structure failed. The crew actuated the escape chute but either didn't have time to use it or were incapacitated when the cockpit pressure change caused by

the chute actuation resulted in front cockpit bulkhead failure. The replacement airplane and its crew narrowly missed the same fate when an investigation of the loss of test data revealed an impending failure of the stabilizer's forward attach structure. A redesign of the stabilizer mounting eliminated the overload problem.

BIS trials began in October 1955 with six A3D-1s to determine if the airplane was acceptable for service use. The main deficiency was that its suitability for high-speed, low-altitude missions was limited by its relatively low structural strength. In order to minimize weight, the specification limit load factor at a combat gross weight of 55,942 pounds was only 2.49, far lower than the usual 6. The 7 December 1955 report concluded that the A3D-1 was not acceptable for delivery to fleet operating units until 18 itemized discrepancies were corrected, the most serious of which were the lack of fuel-dumping provisions, nose wheel shimmy (causing internal damage in the forward fuselage), engine compressor stalls, inadequate braking with the engines shut down, and the inadequate catapult holdback restraint and arresting hook attachment structure. Modifications were subsequently developed for all these problems and aircraft retrofitted.

The tail turret received mixed reviews. According to an early BIS report:

> *The Aero 21B Tail Turret system gave very unsatisfactory results during the preliminary phase of the trials. The Board recommends, due to its questionable effectiveness even when operating properly, that serious consideration be given to removing this turret in future production and utilizing the space thus made available for other needed equipment.[7]*

However, a subsequent report stated:

> *Fighter attacks on the A3D-1 at high altitude usually culminated in a long closing run from directly astern. Simulated gunnery runs from more acute angles off the beam required excellent judgment on the part of the fighter pilot even when no evasive maneuvers were conducted and could often be*

The angled deck and the steam catapult meant that the A3D could be operated at full gross weight even from Essex-class carriers like Shangri-La shown here. Note catapult "bridle" dropping into the water (Jay Miller collection)

foiled entirely by a small change of heading made sharply into the fighter as it approached firing range. The lack of visibility aft and the preponderance of attacks from well astern provide ample justification for retention of the tail turret installation.[8]

The A3D-1 was judged to have met or slightly bettered all of the contract performance guarantees. The empty weight guarantee was missed by only 522 pounds, or 1.5 percent. Considering the degree of difficulty that the weight appeared to present at the beginning of the program, this was a resounding success.

Two different launch configurations were considered for the A3D, tail down and level attitude. With the tail-down setup, the airplane attitude at the end of the catapult stroke approximated the angle of attack for maximum lift at the minimum catapult end speed anticipated. The level launch arrangement pulled the aircraft's nose down, compressing the nose gear strut. At the end of the stroke, the nose rebounded when the catapult bridle released, pitching the nose up. The level launch configuration was selected after shore-based and at-sea testing.

Initial field-carrier landings were accomplished using the low, flat approach with a cut of power for the brief descent to touchdown.

Completion and qualification for at-sea trials was delayed by the failure of the arresting hook attachment structure during one arrested field landing. The first at-sea carrier suitability testing was accomplished in April 1956 aboard *Forrestal* (CV-59). The maximum gross weight on launch was 70,000 pounds. Both LSO-cut landings and mirror-guided landings were made. An angle-of-attack indicator for the pilot and external three-light indicator for the LSO were added for the second series of tests and deemed essential for safe carrier landings. One of the major ship alterations required to accommodate the Skywarrior was a widening of the jet blast deflectors on the catapults to protect personnel, equipment, and aircraft on the deck behind an A3D during launch.

Fielding and Deployment

The Fleet Introduction Program (FIP) began in February 1956. It lasted six weeks, whereupon five of the FIP aircraft were delivered to NAS Jacksonville to VAH-1, the first A3D Squadron to be established.

VAH-1 carrier qualifications were accomplished aboard *Forrestal* in October 1956, although some crews reportedly qualified in November while *Forrestal* was en route to the Mediterranean in

The Mk 5 atomic bomb shown here being delivered to an aircraft carrier's storage facility was one of the A3D's weapons. The tail fins have been removed and strapped to the bomb's sides for transportation. (National Archives 80-G-1026987)

Initially, the sole mission of the A3D was to deliver a nuclear weapon, although crews also trained for delivery of mines and other conventional weapons. The *Forrestal* and equivalent size carriers would deploy with a 12-plane squadron of the Skywarriors. The three Midway-class carriers could accommodate nine A3Ds. The air groups on smaller Essex-class carriers would be augmented with three plane detachments. Two of the Skywarriors would usually be on the flight deck, armed with a nuclear bomb, guarded by Marines, and ready for immediate launch toward a predetermined target. Crews for each ready airplane were standing by, prepared to launch within 15 minutes if so ordered.

Since the A2J program had been canceled and the steam catapult and angled deck made A3D operation feasible from Essex-class carriers, carrier qualification was accomplished on *Bon Homme Richard* (CV-31) in June 1956. While challenging, it proved to be acceptable for fleet operations, with three-plane detachments flying for several years from the six small deck carriers deploying in the Pacific. The first A3D unit to deploy in the Pacific was VAH-2 Detachment Bravo with A3D-2s aboard an Essex-class carrier, *Bon Homme Richard*, which departed to the western Pacific in July 1957.

The Navy showed off the range and speed of its new atomic bomber at every opportunity. In July 1956 a Skywarrior, manned by LCDRs P. Harwood and A. Henson, and Lt. R. Miears, demonstrated the performance capabilities of new carrier-jet attack aircraft with 3,200-mile nonstop, non-refueled flight from Honolulu to Albuquerque, New Mexico, in 5 hours 40 minutes, at an average speed of 570 mph. In September, the commanding officer of Heavy Attack Wing One, CAPT J. T. Blackburn, led a flight of two A3Ds

response to the Suez crisis initiated by the Egyptians seizing operational control of the Suez Canal. The first full deployment was also on *Forrestal* and began in January 1957.

The report on the special weapons phase of the BIS trials was also issued in January 1957. It had been conducted at the U.S. Navy Air Special Weapons Facility at Kirkland Air Force Base in Albuquerque, New Mexico, in 1956. It cleared the A3D to carry the Mk 5 through Mk 8 weapons as well as the Mk 12, Mark 15 mod 0, Mk 18, and Mk 91.

A3Ds originally produced with the tail gun were retrofitted with the defensive electronic countermeasures nose and tail as shown on this VAH-10 A3D. However, the cambered leading edge wing was not retrofitted. (Robert L. Lawson collection)

launched from *Shangri-La* (CV-38) in the Pacific across the U.S., non-stop and unrefueled, to a landing at Jacksonville, Florida.

The ongoing coast-to-coast record flight attempts resulted in an embarrassment in February 1957 when the pilot of brand-new A3D attempted to correct an unsafe landing gear warning once at cruise speed after takeoff by momentarily recycling the landing gear control handle. It proved to be a bad idea. The result was that both main landing gear doors, which had been closed, opened and were carried away along with the door actuators. The loss of utility hydraulic system integrity resulted in one main gear not extending with the emergency system. The pilot and his crew bailed out with the A3D on autopilot and aimed at the Mojave bombing range, where it crash-landed on its own with surprisingly little damage.

This incident didn't stifle the record-setting activity. In March, an A3D-1 Skywarrior, piloted by CDR Dale W. Cox, Jr., broke two transcontinental speed records: one for the round trip Los Angeles to New York and return, in 9 hours 31 minutes 35.4 seconds; and the other for the east-to-west flight against the wind in 5 hours 12 minutes 39.24 seconds.

Like the AJ-1, the 50 production A3D-1s were rushed into service, resulting in the deployable squadrons having to initially deal with most of the shortcomings that had been identified during the BIS trials. Nevertheless, the introduction of the A3Ds resulted in far less carnage than the AJ Savage crews experienced. In spite of the better safety record, however, questions arose about the qualifications of pilots selected to fly the A3D.

This was a carry-over from the selection of pilots for the AJ Savage squadrons. From a pilot's standpoint, the multi-engine AJ was more like a big patrol plane than a carrier-based bomber. Patrol plane pilots were more experienced with its mission, too—single-airplane, long-range, all-weather. On the other hand, pilots assigned to patrol squadrons had few, if any, carrier landings, as there was little cross-fertilization between the land-based and carrier-based components of the Navy. Similarly, few carrier pilots had any multi-engine experience. Overly simplified, the question had been whether it was easier to train a carrier pilot to fly a multi-engine airplane for the all-weather atomic bomb delivery mission or train a patrol plane pilot to land on an aircraft carrier. As it happened, then CAPT John T. Hayward, who led the creation of the heavy attack mission in the Navy, had experience both as a carrier pilot and a patrol bomber pilot. Although he favored the overall experience of the latter, his first squadron had a 50-50 mix of the two.[9] The preponderance of patrol plane pilots didn't seem to have been a problem with the AJ Savage. The A3D, however, added an additional degree of difficulty in carrier landings—higher speed and swept-wing handling qualities.

In August 1957 the Commander of the Sixth Fleet, VADM Charles "Cat" Brown, sent a message to the Commander Naval Air Atlantic that said, in part:

> I am convinced we must begin at once to undo the mischief that has been a long time brewing. As a temporary palliative I recommend insuring that at an early date the senior

officers in each (heavy attack squadron) be former carrier pilots. As a matter of course (this) will filter down and do good. While I think this will help, the only truly effective answer is to begin again at the beginning and from a sound premise, namely the precise reverse of what we are now operating on. We must begin with carrier-qualified pilots and train them in multi-engine aircraft.

Among other things, the result was the assignment of the famously aggressive World War II carrier aviator, CAPT J.D. "Jig Dog" Ramage to command the East Coast heavy attack wing and the reassignment of pilots, particularly in the next squadron that was to deploy, VAH-5. One A3D squadron, VAH-3, was also made the non-deploying training squadron and all the squadrons were co-located at NAS Sanford, Florida, when they were not deployed.

A3D-2

The A3D-1 was followed on the Douglas production line by the A3D-2. First flight occurred in June 1956. Although there were numerous detail changes to fix problems experienced with the -1, the major difference was the substitution of slightly more powerful J57-P-10 engines rated at 10,500 pounds maximum thrust without water injection and with it, 12,400 pounds. The bomb bay was modified to allow for an increase in the number of stores that could be carried and to add provisions for an auxiliary fuel tank. The structure was beefed up to increase the limit load factor to 3.1, still only half of the usual at combat gross weight. Starting with the 49th A3D-2, provisions for inflight refueling probe were added along with other detail changes to the avionics. A total of 164 A3D-2s were built.

One of the flight test A3D-2s crashed in January 1958 during autopilot development testing. During a low-altitude, high-speed run on autopilot, the yaw channel went hard over, followed by the pilot's rudder input in the opposite direction. The reversal overloaded the vertical fin, which failed at the fold line. Both the pilot and flight test engineer were killed. A very careful buildup to the flight condition was made to help confirm the sequence and loading. Flight restrictions on autopilot use resulted.

High-altitude ingress and bombing runs were soon perceived to be too susceptible to detection and interception. During 1959, the A3D squadrons began to train for a high-low-high profile, with the first high segment being the cruise to the edge of the radar detection. The pilot then descended for a low-altitude run in to the target and delivered the bomb with a toss maneuver and wingover recovery. Once clear of defenses, the pilot climbed back to altitude for the return flight.

An early capability added to the A3D was inflight refueling, both as a tanker and a receiver. For the tanker mission, two different drogue systems were evaluated initially. In addition to the now familiar hose system, a folding rigid pipe arrangement was tested. The benefit was that the drogue was positioned farther below the tanker so the receiver would not be in the tanker's downwash. However, it took up

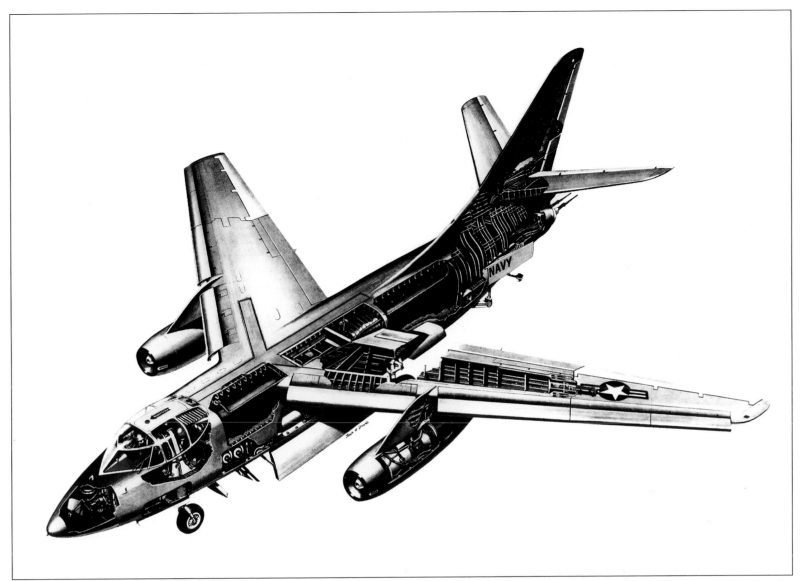

The last 41 A3D-2s were delivered with the cambered leading-edge wing. The change included the addition of a slat between the fuselage and the engine pylon as shown in this cutaway illustration. (U.S. Navy via author's collection)

more space in the bomb bay and did not allow for as much variation in position after hookup between the receiver and the tanker, particularly laterally. Development and qualification of the hose and drogue system was completed in 1958.

As part of an ongoing program to improve the A3D's ceiling, range, and useful load, a cambered leading edge (CLE) wing was developed which also added a leading edge slat between the fuselage and engine pylon. While not all the projected performance was achieved, it did result in a reduced stall speed and a maximum catapult launch weight of 84,000 pounds, which was demonstrated in November 1959 during carrier-suitability tests on *Independence* (CV-62). Nevertheless, for normal operations, the maximum gross weight was limited to 73,000 pounds for carrier launch and 78,000

pounds for field takeoffs. Only the last 41 of the A3D-2s had the CLE wing. They were also delivered with the digital ASB-7 bombing system, which provided better bombing accuracy, and the so-called "dove tail," which substituted ECM protection for the gun turret, which never met expectations. Permanent plumbing for the bomb bay tanker package and the inflight-refueling probe was also installed. The initial, readily removable tanker package reduced the bomb bay capacity from 12 500-pound bombs to eight. Most of these changes, but not the CLE wing, were retrofitted to earlier A3D-2s.

A3Ds produced for missions other than bombing were informally known as "versions." They had a redesigned fuselage with no bomb bay and reinforced canopies and cockpit/cabin structure for

The A3D was used for conventional bombing during the Vietnam War initially, but withdrawn from that mission when the surface-to-air missile defenses became more prevalent. The photo was taken from the jump seat looking across the defensive ECM operator's position, which faced rearward. (Robert L. Lawson collection)

The A3D filled a vital supporting role as a tanker for fighters and bombers prior to ingress and after egress on strike missions in Vietnam. The bombardier/navigator has uncovered the periscope to view the proceedings. (Robert L. Lawson collection)

twice the cabin pressurization differential, which allowed flight at 40,000 feet without oxygen masks. These were the A3D-2P (RA-3B) for photoreconnaissance, the A3D-2Q (EA-3B) for electronic reconnaissance, and the A3D-2T for bombardier/navigator training.

As part of the 1962 redesignation process, the A3D-1 became the A-3A and the A3D-2, the A-3B. Although the A-3Bs were used to drop bombs and lay mines in the Vietnam War beginning in March 1965, the practice was discontinued in late 1966 after SAM installations became more prevalent. The Skywarriors were then assigned to vital supporting roles, first as a dedicated tanker. The designation KA-3B was assigned when the bombers went through the Naval Air Rework Facility at Alameda, California, to receive a semi-permanent tanker package. The KA-3 could transfer about 30,000 pounds of its 43,200 pounds fuel capacity and was deployed in three- to four-airplane detachments.

As the size and weight of avionics decreased, and the radar-aimed threats increased, individual aircraft were increasingly equipped with Electronic Support Measures (ESM) equipment. Initial applications were a simple warning of the reception of the emission of known

The realization of the Navy's late 1940s plan for a super carrier and a long-range atomic bomber are depicted here. Forrestal's first full deployment was made to the Mediterranean from January to July 1957. The air group included F3H-2Ns and FJ-3Ms for fighters and AD-6s, F9F-8Bs, and the A3D-1s for attack. Seven A3D-1s are on deck in this picture with FJ-3s on the axial-deck catapults standing ready. (National Archives 80-GK-22688)

The EKA-3B was the ultimate Vietnam strike support airplane, providing both tanking and jamming services. Note the anachronistic wire antennas still being used on a jet airplane. (Robert L. Lawson)

targeting radar. These became increasingly sophisticated so that they covered more frequencies and provided a directional indication of the threat. Ultimately, they would be capable of limited jamming or deception of the targeting radar to prevent the surface-to-air missile from homing. It was important, however, to not rely too heavily on ESM for self-protection.

When the Vietnam War began, enemy radar jamming was provided by detachments from VAW-33 and VAW-13 flying Douglas EA-1F Skyraiders. It quickly became apparent that the Skyraider was too vulnerable to fly over North Vietnam and its jammers, too low powered to fully protect strike aircraft from offshore. BuAer belatedly issued a requirement for a replacement. In the meantime, the solution was the EKA-3B TACOS—Tanker, Countermeasures, and Strike. The acronym didn't stick and the strike mission wasn't even possible after the conversion, but the Skywarrior was a godsend as a tanker and jammer.

The first five were created from A-3Bs. The next 34 were modifications of KA-3Bs. In addition to the tanker package, the modification added ECM antennas and avionics in pods on both sides of the fuselage, the bomb bay doors, and tip of the vertical fin. Another capability was electronic reconnaissance. The empty weight increased to almost 45,000 pounds with the addition of about two tons of additional avionics and antennas. Since the maximum landing weight

remained at 50,000 pounds, only 5,000 pounds of fuel could be onboard at recovery. As a tanker, the EKA-3B could give about 21,500 pounds of fuel.

VAW-13 received its first EKA-3B in May 1967. The first deployment was aboard *Ranger* (CVA-61) in November. Conversions were made at a fast pace, so within six months there were five three-plane detachments of Electric Whales at sea. The Forrestal-size ships would also have a two-plane detachment of KA-3Bs. The increasing emphasis on the mission resulted in more tactical electronic warfare squadrons being formed from the assets of the former heavy attack squadrons, with VAHs becoming VAQs.

The last EKA-3B detachment returned aboard *Oriskany* in May 1974. However, it was not the last Whale deployment. The electronic reconnaissance version, the EA-3B, flown by VQ-1, VQ-2, and VAQ-33, would continue to operate from the carriers. It was a conversion from the photographic reconnaissance RA-3Bs. A VQ-1 detachment therefore had the honor of the last Whale deployment, which ended in November 1987, with a flyoff from *Ranger*. It was a detachment from VQ-1 flying the electronic reconnaissance version, the EA-3B.

Heinemann said that if he had known that the atomic bomb was going to get smaller as soon as it did, the A3D would have been a lot smaller for the same range. In any event, his next airplane for the Navy took an atomic bomber to the other extreme.

CHAPTER SIX

ONE MAN, ONE BOMB, ONE WAY?

The A4D-2N was the final iteration of the original Skyhawk concept as an atomic bomber, adding a limited all-weather capability. Note the retracted thermal shield above and behind the pilot's head. The control surfaces were painted white to minimize damage by the thermal flash. (Robert L. Lawson collection)

The designers of the first atomic bombs weren't sure how big a yield they would get and in any event were constrained in weight and size by the load carrying and bomb bay size limits of the B-29. The Mk III Fat Man weighed 10,000 pounds; its yield was equivalent to 20 kilotons (kT) of TNT. The Mk I Little Boy was smaller and a little lighter and less powerful. Development soon resulted in bombs of the same size and much higher yields but also smaller, lighter bombs of the same or higher yield that could be carried by tactical aircraft. The first to be fielded, in 1951, was the Mk 8, described in Chapter 2, closely

followed by the Mk 7, which was first issued for service use in 1952. While lighter at 1,680 pounds than the 3,250-pound Mk 8, the Mk 7 was bulky, 15 feet long and 30 inches in diameter, because it used the implosion method of creating a critical mass, the same as the Fat Man. Most of the bomb's outer surface was therefore a fairing ahead of and behind the sphere that was the nuclear device. This made it an awkward load for a fighter or a small single-engine attack airplane. Two other significant advances in the design of the bomb were variable yield, so the explosion could be tailored to the target or the ability of

The Low Altitude Bombing System (LABS) indicator shown here provided the pilot with the guidance necessary to pull up as precisely as possible for an accurate delivery. All he had to do was keep the needles centered until a preset timer released the bomb. (Author's collection)

the delivery airplane to escape the blast, and the remote insertion of the core, which meant that in-flight access to the bomb was no longer required.

The Idiot Loop

The performance and maneuverability of a tactical fighter allowed the use of a delivery technique that avoided two concerns: 1) unwanted attention from the enemy's defenses after a high-altitude approach to the target was detected on radar; and 2) not being far enough away when the bomb exploded. The solution was to make the approach at top speed and treetop altitude, pull up to toss the bomb at the target, and reverse course to depart the other way. Irreverently named the idiot loop, it was developed by the Air Force in 1952 for the delivery of the Mk 7.[1] Three different delivery bomb trajectories were possible, all using the same basic ingress/egress profile—a half-Cuban eight—as shown in Figure 6-1. The entry was from high speed, 500+ knots for jets, and low altitude, 50 to 100 feet off the ground. At a predetermined point, the pilot began a 4 g pull-up, with the bomb automatically released at the proper instant for the selected delivery trajectory. The pilot continued pulling on the stick until he was upside down, headed in the opposite direction of the bomb, and starting to descend. He then rolled right side up and continued the descent back down to 50 to 100 feet, accelerating all the way. When the bomb exploded, the delivery airplane would be five to 10 miles away and rapidly increasing the separation distance.

The Low Altitude Bombing System (LABS) developed in the early 1950s by the Air Force was not essential to performing the maneuver but it reduced pilot workload and improved delivery accuracy. When properly programmed by the pilot prior to takeoff, it indicated when to pull up to toss the bomb, provided guidance to keep him pulling straight up at the right rate, and automatically released the bomb at the right point in the pull up. As shown in Figure 6-1, the bomb would be released at one of three points in the maneuver: low-angle, high-angle, or over-the-shoulder. Each toss point had benefits and shortcomings. The low toss (approximately 45 degrees) provided for the greatest separation from the target itself, which might

mean minimal exposure to enemy anti-aircraft defenses. However, it was the least accurate. The high toss (approximately 65 degrees) had to be made closer to the target, but it provided the maximum bomb hang-time, which meant the greatest separation from the detonation. The over-the-shoulder toss (approximately 110 degrees) increased the risk of being shot down as well as that of being blown up, but it was sometimes necessary because of the need to attack a target for which a satisfactory initial point did not exist. It was also the most accurate and provided surety of target identification.

Low-altitude ingress minimized detectability but it made navigating to the target and attacking it more difficult. Like an instrument approach, the delivery technique required an initial point (IP), which was a prominent and readily identifiable landmark—such as a bridge or a geographic feature—within a few miles of the target. It had to be visible from a distance even at a height of 50 feet, because the pilot had to fly over it while headed toward the target, which would be out of sight.

Over the IP, the pilot would start the preset LABS countdown by pressing the bomb release "pickle" button on the stick grip. A tone indicated that the countdown was in progress. When it stopped, the pilot was to immediately begin the pull up, following pitch and heading command needles on his LABS indicator to insure that he was pulling up vertically and at the right rate. The LABS would release the bomb at the point where it would follow a ballistic trajectory to the target.

If a suitable lead-in IP was not available, then the target itself had to be the initial point. This had two advantages: It was the most accurate and the target could be attacked from any direction, terrain permitting, without regard to the availability of landmarks for an initial point. However, it exposed the pilot to air defenses around the target both coming and going. Since the target had been over flown before the pull-up for the delivery, the bomb had to be released *after* the airplane had passed vertical in the half-Cuban eight, hence the name over-the-shoulder. There was less separation from the detonation than in a toss because the bomb and the airplane were traveling in same direction after its release, albeit with the bomb going mostly up while the airplane was going down and out. It was therefore the maneuver most likely to require reduction in weapon yield and/or burst height.

Another low-altitude delivery option developed later was the laydown, which was done at high speed while overflying the target; the bomb being slowed by a parachute or other drag device while the aircraft departed the premises.

In addition to LABS, a weapon control box was added to airplanes that were to be used for atomic bomb delivery. This provided for the reconfiguration of the bomb's tail fin if required, the arming of the bomb (insertion/extraction of the core), and verification that it was armed and its control circuits were functioning properly.

Interim Tactical Nuclear Bombers

Several different single-seat fighters and bombers were modified for delivery of atomic weapons and used operationally while the A4D Skyhawk and subsequently the FJ-4B Fury were in development.

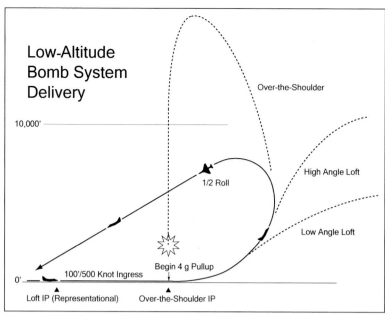

Figure 6-1. (Author)

Douglas AD Skyraider

The AD Skyraider was a significant part of the Navy's capability to deliver the atomic bomb for many years. What it lacked in speed, it made up for in range, albeit by being able to fly for what seemed like forever. That range could also be achieved at low altitude, making it valuable for air defense suppression prior to the ingress of the jets, which preferred to cruise as long as possible at the altitudes that would provide the enemy with early warning. The single-seat AD-4s were replaced by the AD-6 beginning in late 1953. These had centerline and inboard wing pylons that could accommodate heavier and/or larger loads than those on the AD-4 and some strengthening of the airframe structure to permit higher maneuver load factors and higher sink rate landings.[2] The maximum gross weight was not increased however. Cockpit changes included the addition of an engine torque-pressure gauge for improved power management and Tactical Air Navigation (TACAN) and a UHF homer for better navigation. A LABS control panel replaced the APS-19 radarscope, which was no longer to be carried.

The special weapons mission profile flown by the AD-6/7s was known as Sandblower. After launch, the pilot would proceed to the target at an altitude of 50 to 100 feet, approximately the height of the carrier deck above the water. All navigation over the water was dead reckoning, compensating for wind by determining its direction and velocity by the wave top action. Navigation once ashore was augmented by whatever landmarks could be seen and identified from treetop altitude. At the takeoff gross weight of 22,000 pounds with full internal fuel and two 300-gallon drop tanks, maximum range cruise speed was about 140 knots, increasing to 170 knots as the weight of fuel decreased. Missions would therefore be as much as 12 to 13 hours in duration with the mission radius approaching 1,000 nm.[3] Some 6th Fleet targets, such as those in Romania and the Ukraine, exceeded the range of even the

AD, so there were designated refueling airstrips in friendly countries to be used during the return. Pacific Fleet pilots, with targets in southeastern China and far eastern Russia, faced the same challenge.

Of course, the AD couldn't toss the bomb quite as high or far as the jets because even at full throttle with water injection, it was only going to be running in at 275 knots, a little better than half the speed of a jet. Because faster and lighter was better, the preparation for the special weapons delivery mission included the removal of the external armor, the four 20mm cannons, and the stores pylons on the outboard wing panels. The resulting holes were covered by what later became known as duct tape.

Skyraider pilots were grateful when the weapons engineers at the Naval Ordnance Test Station (NOTS) added a rocket motor to the Mk 7 as shown in Figure 6-2 so that it traveled much farther after being lofted at the target. This was known as the BOAR, a 30.5-inch Rocket, Mk 1 Mod 0. BOAR, depending on who's telling the story, stood for either Bombardment Aircraft Rocket or Bureau of Ordnance Atomic Rocket, with the former seeming more credible. It had a maximum range of about seven miles when fired at a toss angle of about 20 degrees, which made the recovery after the launch a lot less dramatic than the higher-angle tosses. Development began in 1952 at NOTS with the first flight test in June 1953. It was operational in 1956. BOAR practice missions were accomplished using a 5-inch rocket to simulate the trajectory of the weapon.

Douglas built a total of 713 AD-6s. These were followed by only 72 of the almost identical AD-7s, with the last one delivered in February 1957. (An additional 168 were canceled because replacement of the Skyraiders with the A4D Skyhawk had begun in 1956.) Nevertheless, the Navy continued to deploy Skyraiders through the early part of the Vietnam War. Combat highlights included the shooting down of two MiG-17s, one on 20 June 1965 by two pilots from VA-25 and another on 9 October 1966 by Lt. (j.g.)William T. Patton of VA-176. However, the propeller-driven Skyraiders lacked the performance to keep up with the rest of the air wing's strike group.

The transition took a decade, with VA-25 being the last attack squadron deploying the Skyraiders, now designated A-1H and A-1J for the AD-6 and AD-7, respectively. For that last combat tour with Air Wing 15 aboard *Coral Sea* from July 1967 to April 1968, they got A-1s modified with the Stanley Aviation Corp. extraction seat to provide the pilot with an improved escape capability. VA-25's last combat mission was flown in A-1H (BuNo 135300), which is now on display in the National Museum of Naval Aviation in Pensacola.

However, the single-seat Skyraider continued to soldier on with the U.S. Air Force and the Vietnamese Air Force because of its unparalleled capability—endurance and ordnance load—in close air support.

Jet Fighter Placeholders

The Navy, like the Air Force, initially modified existing jet fighters to carry the tactical nuclear weapons. The F2H-2B was the first, with B being the designation for different armament. (The B in F9F-2B represented the addition of bomb and rocket pylons, not a nuclear delivery capability.) While now no heavier than a World War II

The blue shape beneath the fuselage of this AD-6 is a practice Mk 7. The pilot extended the lower tail fin after takeoff. The Skyraider has not been completely prepared for a mission because it still has the outboard wing stores pylons and 20mm cannons. (Robert L. Lawson collection)

When the AD Skyraider first deployed, the front-line Navy fighter was the propeller-driven F8F Bearcat shown here. It would serve long past the transition to jet fighters and even the introduction of jet attack planes. (Robert L. Lawson collection)

On the AD (now A-1) Skyraider's last deployment aboard Coral Sea with VA-25 in 1968, the front-line Navy fighter was the Mach 2 F-4B Phantom. This squadron was one of two flying A-1s with the Stanley Aviation Corporation extraction seat. (National Archives 80-GK-43788 via Ed Barthelmes)

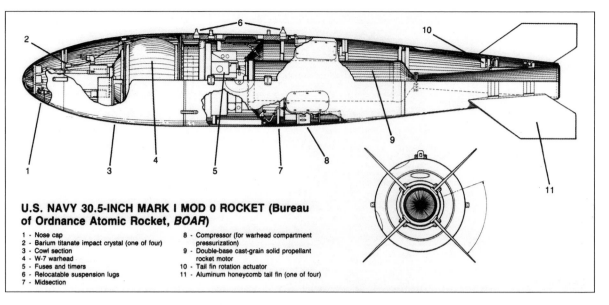

Figure 6-2: In order to provide greater separation from the Mk 7's atomic blast as well as the target's defenses, a rocket motor was added to create an unguided atomic rocket. (Jay Miller collection)

U.S. NAVY 30.5-INCH MARK I MOD 0 ROCKET (Bureau of Ordnance Atomic Rocket, BOAR)

1 - Nose cap
2 - Barium titanate impact crystal (one of four)
3 - Cowl section
4 - W-7 warhead
5 - Fuses and timers
6 - Relocatable suspension lugs
7 - Midsection
8 - Compressor (for warhead compartment pressurization)
9 - Double-base cast-grain solid propellant rocket motor
10 - Tail fin rotation actuator
11 - Aluminum honeycomb tail fin (one of four)

The F3H Demon, like all Navy jet fighters at the time, was expected to accomplish strike missions as a general-purpose fighter. This VX-4 Demon is armed with an odd combination of one very big, low drag 3,000-lb AN-M118 "demolition bomb" and six AN-M30A1 100-lb bombs. (Terry Panopalis collection)

torpedo, the atomic bombs were still very heavy when compared to the modest payload capability of the early carrier-based jets. On *Bon Homme Richard* in late 1951 during the Korean War, the F9F-2 Panther could be armed with up to six rockets weighing a total of 840 pounds (in addition to its full complement of 20mm cannon ammunition) but only if the wind-over-deck exceeded 30 knots. For every knot of wind-over-deck more than that, two rockets could be added.

Below 30 knots Wind Over Deck (WOD), the Panthers stayed on deck. That was not necessarily a problem even if there was no natural wind because with a clean bottom and all boilers on line the Essex-class carriers could generate up to 33 knots.

The first F2H-2B deployment was as a VC-4 detachment aboard *Coral Sea* in April 1952. Mission radius was initially restricted because lateral cg limitations for roll control immediately after a

This Banshee, BuNo 125067, is one of 27 F2H-2Bs modified to deliver either the Mk 7 or the Mk 8 atomic bomb. The modification included the addition of a large stores pylon shown here just outboard of the engine inlet and an inflight refueling probe, barely visible here, extending out of the right outboard 20mm gun port. In 1955, this F2H-2B was assigned to VF-101. (U.S. Navy via author's collection)

The size of even the tactical nuclear weapons made for awkward situations. With its main landing gear struts held extended with a metal sleeve, the F2H could taxi (very carefully) and takeoff with the Mk 7 (in this case in the BOAR configuration) on a one-way trip. The sleeves dropped out after takeoff during retraction of the landing gear. (U.S. Navy via author's collection)

catapult launch did not allow filling the tip tank on the bomb side of the airplane and the gross weight limit did not permit adding a compensating weight on the other side. Serendipitously, a British company was developing in-flight refueling in the late 1940s. The Navy evaluated and quickly adopted the probe and drogue inflight refueling system because it allowed an airplane to be catapulted off with the heavy weapon and reduced fuel load, climb to altitude, and then have its tanks topped off. Mission radius was further increased if the jet bomber could be refueled on the way back.

With inflight refueling, the F2H-2Bs had strategic range. Whitey Feightner participated in an exercise in 1953 that began with a launch of three Banshees from *Midway* cruising off Cuba. After refueling on the way north by AJ Savages, they went low level to Lake Erie until they were detected by Air Force radar. They eluded the interception by climbing to 50,000 feet, dropped their "shapes" into Lake Erie, and returned to Guantanamo, Cuba, after another inflight refueling. It was a round trip of more than 2,800 miles.[4]

The F2H design was stretched and otherwise modified to create the F2H-3, and with a radar system change, the F2H-4. The nuclear-capable Banshees were initially assigned to VC squadrons, which supplied detachments to the aircraft carriers. The follow-on to the F2H-2B, the F2H-3/4, was stipulated to have nuclear weapon delivery capability from the beginning, so it did not have a B suffix.

Another interim type to fill in until the A4D became operational was the F9F-8B. It was modified and qualified to carry and deliver a nuclear weapon and assigned to VA squadrons as were F7U Cutlasses for a short time.

The Bantam Bomber

Ed Heinemann had been working on a lightweight fighter study in the early 1950s that was redundant to forthcoming Navy contracts at North American (FJ-4) and Grumman (F11F).[5] The Navy requested that he instead propose a light jet attack airplane, also powered by the Wright J65, for delivery of a nuclear weapon. Their expectation was 500 knots maximum speed, 400 nm minimum radius of action on internal fuel, and a gross weight of 30,000 pounds.

Douglas' subsequent unsolicited proposal was the basis for a contract in June 1952 for a single XA4D using funds from the troubled A2D program. (The A4D Skyhawk wasn't a substitute for the A2D in the Navy but it was a substitute for the A2D at Douglas.) Based on innovative design work and the active participation of the BuAer class desk, CDR John Brown, the little airplane was to have a gross weight of only 15,000 pounds. The mockup review was accomplished in October 1952. Building on his success with the AD Skyraider, Heinemann outdid himself with the A4D, stripping the airplane down to its essentials. The cockpit was as small as possible. The ejection seat and pilot's restraint and parachute harness were redesigned to minimize weight. The avionics were minimal because the A4D was to be a "day-visual attack airplane" and to further reduce weight; the components were removed from the individual boxes and integrated into one box, eliminating multiple power supplies and enclosures for a start. The wing was a clipped delta with the wingspan limited to the 27.5-ft maximum allowed for a folded airplane so the weight of a folding system and structure was not required. The usual armament of four 20mm cannons was reduced to

A VC-5 AJ-2 tanker tops up a VF-31 F2H-3 Banshee carrying a practice Mk 7 nuclear bomb, circa 1954. (Robert L. Lawson collection)

The smaller Mk 12 atomic bomb shown here on an F9F-8 did not have the ground clearance problem of the Mk 7. The large tail fins folded to the side to clear both the extended flaps and the ground and were deployed after takeoff. (U.S. Navy via Hal Andrews collection)

The Mk 12's tail fins—shown folded here on an F9F-8 for ground and extended flap clearance—had a kink on the leading edge to produce a football-like spiral after release, reducing variation in the bomb's ballistic trajectory. (U.S. Navy via Hal Andrews collection)

The A4D mockup clearly depicted the extra length of its landing gear, which provided more adequate ground clearance for all the atomic bombs. The engine inlet and exhaust would change during detail design of the prototype. The canopy would also be changed to a conventional two-piece arrangement, although the prototypes would not initially have a separately framed windshield. (Gary Verver collection)

two. Inflight refueling capability was originally not incorporated. Vendors were given a challenging weight number to meet; failure to do so was not acceptable.

One of the few concessions to weight but consistent with the requirement to carry an Mk 7 was the length of the landing gear, which also required a waiver of the usual turnover angle limit. Adapting existing aircraft to carry the Mk 7 had required a folding fin on the bomb and the acceptance of minimum ground clearance. The A4D could even land carrying the Mk 7. There were also only three stores pylons, a centerline station for the bomb and two wing stations for external tanks that were to be dropped when empty.

Not only did the program sidestep any competition, it quickly escalated into an order for 20 production aircraft including the original XA4D and a static test article. The customary initial order was for two experimental flight test aircraft and a static test article to be followed in due course by a smaller, low rate production order. If all went well initially in flight-test, a higher rate production order would result. The Navy clearly wanted to take advantage of the reduction in size of nuclear weapons as soon as possible.

The XA4D flew for the first time in June 1954, closely followed by the first production A4D-1 in July. Flight test, proceeding concurrently with low-rate production, was not trouble-free, but progressed quickly compared to the F4D Skyray that Douglas was developing at the same time. However, Jimmy Verdin, who as a Navy officer had flown an XF4D to a world speed record in 1953, was killed in the

The A4D's cockpit was simple and uncluttered. The striped panel on the left was pulled to jettison all stores; the one on the right, to shut off the hydraulic pressure to the flight controls, which required extending the control stick for more leverage. The instrument panel eventually extended downward between the pilot's knees as switches had to be added. (Gary Verver collection)

An example of A4D weight reduction was the redesign of the Douglas ejection seat used in the A2D and the F4D. Although slightly misrepresented (the Skyhawk seat that the Wave is lifting does not have a seat belt or the shoulder straps that connected to the pilot's torso harness), there was a significant difference. (Gary Verver collection)

This photo of an early A4D-1 (it has the one-piece windscreen and no vortex generators on the wing slats) shows the increased ground clearance of the Mk 7 shape compared to that on the F2H Banshee. (Gary Verver collection)

Carrier qualification of the A4D-1 in September 1955 went smoothly. For some reason, the aerodynamically actuated leading edge slats are retracted on this Skyhawk, which is probably taking a wave-off. (Gary Verver collection)

There was a price to be paid for stores ground clearance and stubby wings and it was poor turnover angle and cross wind capability. This A4D-2 has fallen and can't get up aboard Midway in November 1960. (U.S. Navy via author's collection)

crash of an A4D in January 1955 while evaluating a fix for transonic wing drop. Following the failure of the single hydraulic control system, he was forced to eject at high speed and was probably knocked unconscious by the canopy. The successful fix for the wing drop was the addition of two rows of vortex generators on the outboard wing panel forward of the ailerons.[6]

Initial shipboard trials were accomplished aboard *Ticonderoga* (CV-14) in September 1955. The report included an unusual commentary:

From the carrier suitability standpoint, the general design philosophy embodied in the A4D-1 has produced excellent results. Structurally, the airplane has few shortcomings of import and only in the arresting hook installation has there been indications of surplus strength and the extra weight thereby implied. The reduction in overall airplane complexity has contributed to excellent availability.

In October 1955, Lt. Gordon Gray increased the 500-km closed-course speed record to 695 mph (604 knots), one of the very few times a speed record was held by an attack airplane as opposed to a fighter.

BIS trials began in March 1956 with six aircraft. As part of this activity, the Naval Air Special Weapons Facility at Kirtland Air Force Base, Albuquerque, New Mexico, accomplished a nuclear weapons evaluation during April, May, and June. The overall compatibility with the Mk 7, Mk 12, and Mk 91 weapons was judged to be satisfactory. (The Mk 91 replaced the Mk 8) The Mk 12 could even be brought back with the fins extended.

The BIS report on the A4D-1 dated 11 December 1957 "cited eighteen mandatory for-correction items and a number of items rec-

ommended for corrective action. The Chief, Bureau of Aeronautics, has advised that corrective action is being taken where economically feasible, and that uncorrected deficiencies do not prevent the airplane from performing its mission. The only serious uncorrected mandatory deficiency is a failure to meet specifications for wind-over-deck required for launch on the H-8 catapult. Several undesirable handling qualities are inherent in the design of the airplane; i.e., cross-wind landing characteristics, low maneuver capability at high subsonic speeds, low buffet boundary, weak lateral and longitudinal stability in cruising flight."

The cover letter went on to state that the A4D-1, the last of which had been accepted months earlier in July 1957, failed to meet guarantees for takeoff distance and combat ceiling but "Later production contracts contain less stringent guarantees, and the airplane will meet these." It was good to be a favored Navy supplier.

In the meantime, the Fleet Indoctrination Program had begun at VA-72 on the East Coast and VF(AW)-3 on the West Coast in late 1956. In October 1956, VA-72 at NAS Quonset Point, Rhode Island, was the first squadron to take delivery of the A4D-1 from the factory. Only 165 were built. VA-93 on the West Coast became the first squadron to deploy with the A4D-1 aboard *Ticonderoga* in September 1957. The A4D-2, which followed the -1 down the production line, had inflight refueling along with other structural and component improvements and Bullpup delivery capability. 542 were produced with first deliveries in late 1957.

The FJ-4B

Concurrently with the Douglas A4D program, North American was developing the FJ-4, also powered by the J65. Initiated as a

Inflight refueling greatly extended the A4D's mission radius. Here a VX-1 A4D-2 carrying an Mk 7 is being refueled in flight from another A4D equipped with the Douglas developed "buddy" store, which contained a hose reel, pump, and fuel. Fuel could also be transferred from the tanker's external and internal tanks to the buddy store. (U.S. Navy via T. Scott collection)

lightweight, highly maneuverable, low-cost fighter, it did not have an afterburner. Although it met specification requirements for speed, range, and turning capability at altitude, the Navy fighter community rejected it in favor of the supersonic F8U Crusader. The production airplanes were all delivered to the Marine Corps. BuAer, however, in keeping with its inclination to have a low-risk backup for a new airplane program, placed a contract with North American for the FJ-4B in July 1954, the month after the Skyhawk's first flight. The additions were minimal: two more stores stations, another pair of speed brakes (the original pair of brakes were not effective enough), spoilers to increase roll-control authority (and counteract the weight of the Mk 7 on an outboard pylon), and LABS. First flight was in December 1956, just after first deliveries of the A4D-1 for FIP had begun.

North American also produced a unique in-flight refueling pod for the FJ-4B. The tanker would carry two similar external stores, one the refueling pod and the other a 200-gallon external tank with a self-contained, air-driven pump. The bomber would take off with the nuclear weapon and three external tanks and be refueled in flight by the tanker, providing a combat radius of 1,115 nm with a 2,395-pound Mk 28 store. CAPT William C. "Bawldy" Chapman, USN(ret), remembers that he

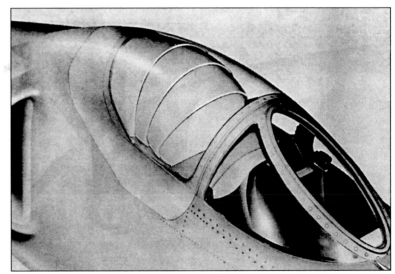

The most difficult atomic bomb effect to outrun was the thermal flash. Bombers were therefore equipped with thermal protection devices that quickly closed off the cockpit following release of the bomb. (U.S. Navy via author's collection)

To compensate for the Skyhawk's limited navigation capability and augment its mission radius, one operational concept was to have an A3D lead two A4Ds to an en route waypoint and refuel them prior to each aircraft proceeding independently to its assigned target. The refueling reduced the A3D's mission radius but its maximum range was only required for a subset of the Navy's SIOP targets. (Robert L. Lawson collection)

North American also created a buddy system, in this case specifically for the FJ-4B. The refueling drogue and hose reel were contained in the pod under the right wing. Both the refueling tank and the large external tank under the left wing were equipped with a wind-driven turbine on its nose to provide power to the fuel pump and hose reel motor. Any FJ-4B could be configured as either a bomber or a tanker. (Robert L. Lawson collection)

The FJ-4B's six stores stations allowed it to carry five Bullpups along with a pod containing the guidance avionics. Because it only had three pylons, the early A4D could carry a maximum of three. It would have been just two if the guidance equipment had not been somehow stuffed into the airframe. (Robert L. Lawson collection)

One of the relatively few squadrons to deploy with the A4D-1 was VA-113. The A4D-2 with inflight refueling capability was already in development. (Robert L. Lawson collection)

and his executive officer were prepared to launch from their carrier in 1958 off Okinawa against separate targets in China, with his exec landing in Calcutta and himself at Shemya in the Aleutians.[7]

Carrying an Mk 28 store and as many external tanks as possible, the two aircraft were similar in performance for a sea-level delivery mission according to their respective SAC charts dated 1 July 1967.

	A4D-2	FJ-4B
Takeoff Weight (lb)	22,130	26,893
Fuel Including Max External (lb)	9,530	9,112
Mk 28	2,025	2,395
Mission Radius (nm)	550	565
With Inflight Refueling (nm)	890	1,115
Combat Weight (lb)	15,625	23,187
Sea Level		
Maximum Speed (kt)	562	542
Rate of Climb (fpm)	7,800	5,100
Combat Ceiling (ft)	41,300	40,000

Although the FJ-4B was arguably superior to the A4D in some respects, such as having twice as many stores pylons, only 222 were

As additional protection against thermal flash, pilots were issued the all-white anti-flash suit. Wynn Foster, better known as CAPT. Hook, models it here while flying with VA-76 in the early 1960s. (CAPT Wynn Foster, USN, retired)

The FJ-4B could carry the Mk 7 as well as three external tanks, not shown here, giving it a range advantage over the A4D. The extra pair of speed brakes is just visible as a ribbed panel aft of the VA-216 marking. (Hal Andrews collection)

produced, the last being delivered in May 1958. The preference for the A4D appears to have been lower unit cost and maintenance man-hours. For convenience in support and training, all the FJ-4Bs were assigned to Pacific Fleet squadrons. The last one, VA-216, turned theirs in and received Skyhawks in 1962.[8]

Size Matters

Even before the A4D and FJ-4B began to deploy, some questioned the value of the A3D versus the smaller airplanes. There's only so much space aboard an aircraft carrier. Smaller is better, all other things being equal. Various metrics have been used to compare the relative

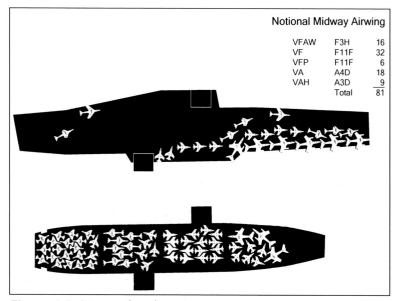

Notional Midway Airwing		
VFAW	F3H	16
VF	F11F	32
VFP	F11F	6
VA	A4D	18
VAH	A3D	9
	Total	81

Figure 6-3. (National Archives)

space that an aircraft type takes up on the carrier. Immediately after World War II, it was the number that could be spotted on the first 200 feet of an Essex-class carrier. In the mid-1950s, it was the maximum number that "can be accommodated in a landing spot on the flight and hangar decks of a CV-19 class angled-deck carrier." Later, it became a ratio of a baseline aircraft, which allowed a rough calculation to be made of the number of a mixed fleet that could be accommodated.

Figure 6-3 illustrates the relative space required on a Midway-class carrier for the A3D and A4D, with the deck spotted for recovery as well as keeping two catapults open for alert fighter launches. By the maximum number method, 106 (with the refueling probe) of the diminutive A4Ds, even though it didn't fold, could be parked on a CV-19. There was room for 78 FJ-4Bs and only 25 A3Ds. The trade-off, of course, was that the A3D could carry a 10,000-pound bomb load out 1,200 nm and return whereas the A4D could only manage 2,000 pounds to 800 nm. With ingress refueling, an FJ-4B could deliver an Mk 28 at roughly the same mission radius as an A3D. Using a buddy tanker, of course, could effectively cut the number of bomb-carrying airplanes in half. The Skywarrior was also capable of radar navigation and bombing, whereas the A4D and FJ-4B pilots were ill equipped to reach, much less find, their targets at night or in bad weather. On the other hand, light attack proponents argued that the little jets were more likely to penetrate and survive enemy air defenses.

A4D Improvements

Douglas was on contract in 1957 for the A4D-3. It was to be powered by the Pratt & Whitney J52, which provided the same thrust as the Wright J65, but at lower fuel consumption. An all-weather capability was to be provided for by the addition of APG-53 radar (in a 9-inch longer nose) that would provide terrain clearance and navigation capability along with slant range data for a Mk 9 bomb director for toss bombing, an autopilot, an all-attitude indicating gyro, and a windshield rain removal system. The leading edge of the wing was to be cambered since the A4D-1/2 wing slats had initially been troublesome in service, opening and closing asymmetrically. However, the Navy decided not to invest in a new engine or the modified wing, limiting the upgrade to the avionics that provided all-weather capability. The result was designated the A4D-2N with an empty weight increase of about 600 pounds. 638 were produced to replace the A4D-2s and FJ-4Bs.

The Navy's procurement of the FJ-4B may have spurred Douglas to propose the A4D-4 in 1958, also powered by the J52. It was a significant configuration change, with the wing becoming swept and greater in span, requiring folding. Four more hardpoints were added, two on each wing. The fuselage was lengthened to provide more volume for fuel and electronics. The rearward vision was improved with the change to a bubble canopy. Navigation was to be enhanced with a dead-reckoning navigation computer using inputs from an astrotracker and Doppler radar. The Navy again passed, probably due to budget limitations.

The lack of A4D stores stations for conventional weapons delivery was a significant limitation. Because of the need for external tanks, the

The military aircraft manufacturer's business model relies on the frequent generation of contracts for new or improved airplanes that will replace existing ones. Most proposals go unrealized, like this J52-powered A4D-4, paying even less than the usual lip service to being a modification of an existing design. (U.S. Navy via author's collection)

The bomb rack adapter allowed the A4D to be more productive performing conventional weapons-delivery missions. With only three stores stations, the maximum number of bombs the A4D could carry was three at some penalty in range, since this precluded the use of external tanks. A greater number of small bombs provided more close air support capability than two or three large ones, and increased to the likelihood that one would hit the target. The configuration in this publicity picture includes two bombs that are clearly interfering with the main landing gear doors. (U.S. Navy via author's collection)

A4D-1 and -2 could only carry one bomb with two external tanks or two bombs with one external tank. VX-5 at China Lake came up with a solution in 1959, the Multiple Carriage Bomb Rack (MCBR). The prototype used six of the small Aero-15 racks from a crashed AD Skyraider welded to an adapter that was hung from the standard A4D rack. It helped that the A4D landing gear provided plenty of ground clearance. The first bombs, Mk 81s, were dropped from the adapter in October 1959. In December, a VX-5 A4D carried and dropped 16 Mk 81s using adapters on all three A4D pylons (only five bombs could be carried on the wing pylons because of the location of the landing gear doors) at an MCAS Yuma, Arizona, firepower demonstrator for the Deputy Chief of Naval Operations for Air Warfare, VADM Pirie, who immediately appreciated the value of the concept. In short order, Douglas submitted an unsolicited proposal for Multiple Bomb Racks and was awarded a contract in early 1960. The first production racks were delivered and qualified by June.

The prototype MCBR and the Douglas MBR simply released the bombs. The next refinement to the concept was the use of cartridges to provide positive separation of the bomb from the rack. This was the Multiple Ejector Rack (MER). For some applications, six bombs were too many so the shorter Triple Ejector Rack (TER) was created that only carried three bombs per pylon. The MER and the TER eliminated the need for multiple pylons on outboard wing panels, which were hard to load on aircraft with folding wings. The only shortcoming was the increase in weight and drag of the MER/TER that had to be taken into account in mission planning.

Two additional wing stores stations were finally added to the Skyhawk with the A4D-5, which was ordered in July 1959 with a first flight in July 1961. The engine was finally changed to a Pratt & Whitney J52-P-6, providing an empty weight reduction, an 800-pound thrust increase, and lower fuel consumption compared to the Wright J65. The engine change was combined with an inlet change to add a

The Mk 4 gun pod was developed to augment strafing capability when desired. It contained a Mark 11 20mm gun that was self-powered and twin-barreled. A maximum load of ammunition was 750 rounds, which could be fired at a maximum rate of almost 70 rounds per second. Development tests were completed in 1963. It appears to have been rarely used by carrier-based airplanes, reportedly because it was unreliable. (U.S. Navy via author's collection)

The A4D-5 was the final step in the transition of the Skyhawk to a more useful aircraft for conventional warfare. The J52 engine provided more thrust, and two additional stores pylons provided more bomb attach points. (Robert L. Lawson collection)

splitter plate. The nose was lengthened again to provide volume for more avionics, with miniaturization helping the cause. New mission functionality included TACAN, Doppler navigation, Mk 9 toss bombing system, a radio altimeter, and an updated LABS. The A4D-5 replaced the A4D-2N in the fleet beginning in December 1962, with 500 produced.

In 1962, the A4D-2N was redesignated the A-4C and the A4D-5, A-4E.[9] In 1964, when the Vietnam War began with the Gulf of Tonkin incident in August, the Skyhawk was the only light jet attack aircraft in carrier air wings since the FJ-4Bs had been retired. The first loss occurred almost immediately, when LT(jg) Everett Alvarez was shot down on 5 August by AAA fire to become the first Navy prisoner of war. Developed primarily to deliver an atomic bomb, the Skyhawk finished its combat career in the Navy as a tactical bomber and air-defense suppressor, although it served for many more years doing an excellent MiG-17 imitation in air-to-air combat training of Navy fighter pilots and serving with the Navy's Blue Angels demonstration team.

One of the major changes based on combat experience was the addition of defensive avionics to detect enemy radar activity, especially the targeting of Soviet-furnished surface-to-air missiles (SAMs). Designed to knock down big, lumbering bombers, a SAM could be defeated by a last moment sharp turn that it couldn't match. However, the pilot had to see it in time to get set up for the maneuver. Also, having to deal with two or more SAMs at the same time greatly increased the degree of difficulty. In August 1965, a program was initiated with Douglas and Sanders Associates to adapt the Sanders ALQ-51 E-F band deception equipment to the A-4 Skyhawk. Installing all the black boxes, antennas, and wiring was accomplished under the aptly named *Project Shoehorn*. The first modified Skyhawks deployed on *Constellation* (CV-64) in October 1965. According to Sanders advertisements, the effectiveness of the SA-2 was one kill for 10 missiles fired at unprotected aircraft but only one for 50 with countermeasures installed.

Another self-defense retrofit was the ALE-29A Countermeasures Dispensing Set. This self-contained box ejected decoy flares or chaff

The Catapult Officer is giving the signal to launch this VA-12 A-4E on 6 September 1966. It is fitted with Shoehorn, as indicated by the ice cream cones pointed forward and aft, and armed with 250-lb bombs, Snakeyes on the center pylon, and World War II leftover box-tail M57s on the outboard pylons. (National Archives 428-K-33067 via Angelo Romano)

as selected by the pilot. Each dispenser could be loaded with 30 cartridges, and two dispensers were usually installed in the underside of the aft fuselage. The flares provided an alternate target for an incoming heat-seeking missile while the chaff was somewhat effective against radar used to aim antiaircraft guns or SAMs.

Zuni Rocket

The World War II unguided 5-inch high velocity aircraft rocket (HVAR), which could only be mounted one to a pylon, was replaced by the Zuni folding fin rocket, which came four to a pod. This

These VA-212 A-4Es are taxiing for takeoff from Bon Homme Richard in 1967, armed with at least one of almost everything. Rampant Raider 227 is armed with a Shrike and a Bullpup (with what appears to be a modification to its nose), 233 is carrying a Walleye, and 221 is loaded with a Shrike and a Triple Ejector Rack (TER) with Snakeyes. (Robert L. Lawson collection)

Dick Stratton had this to say about the accuracy of the 2.75-inch folding-fin rocket: "The Aero 7D Rocket pack (had) 19 independently targeted warheads, the destination of which even I did not know." His opinion may have been colored by an unfortunate collision and explosion of a couple during an attack on a North Vietnamese bridge in January 1967. The engine of his A-4 ingested some of the debris and subsequently failed catastrophically, resulting in him being a POW for the duration of the war. (U.S. Navy via Gary Verver)

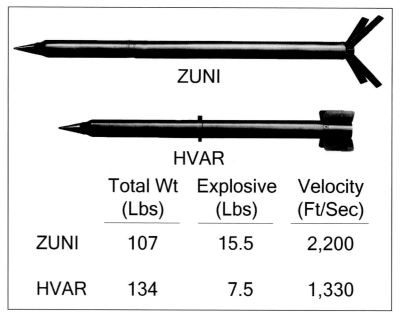

	Total Wt (Lbs)	Explosive (Lbs)	Velocity (Ft/Sec)
ZUNI	107	15.5	2,200
HVAR	134	7.5	1,330

The Zuni was the replacement for the 5-inch High Velocity Aircraft Rocket (HVAR). The fins on the Zuni were spring-loaded to the extended position. With the fins folded back, the Zuni was loaded in a 5-inch tube in a 4-rocket pod that quadrupled the number of rockets that could be carried on a single stores pylon. (Author)

allowed a quadrupling of the number of rockets that could be carried for the same weight since a 27-pound reduction in the weight per rocket compensated for the pod weight. It was developed at the Naval Ordnance Test Station, China Lake in the early 1950s as a bigger brother of the Mighty Mouse, a 2.75-inch folding fin rocket originally developed as an air-to-air weapon to be used by interceptors against heavy bombers. It was approved for production in 1957. Various warheads could be accommodated for anti-armor, delayed-action, proximity-fused, etc. Depending on the warhead, the Zuni was about 110 inches long and weighed 107 pounds. Accuracy depended on whether all the fins unfolded simultaneously; but if they did, the Zuni could be a bit more accurate than the HVAR because of its higher velocity.

Bullpup

The Zuni delivered a devastating blow in close-air support but it lacked accuracy. A new guided missile, Bullpup, was developed along with the A4D-2 and the FJ-4B and first deployed in 1959. Rocket propelled, it was guided by the launch pilot, who moved a small joystick or control button to keep the tracking flare burning on the back of the missile superimposed on top of the target. This meant that the pilot had to follow along behind it with minimal maneuvering until missile impact. The Air Force employed the Bullpup missile as well although as usual, the details, designation, and even the popular name were initially different. The Navy's ASM-N-7A (later AGM-12B) was also known as Bullpup A. It was rail launched, had a 250-pound warhead, and weighed 570 pounds with a range of three to six miles. Recognizing that its warhead was a little small, even with a speed at impact of Mach 2, the Navy developed the ASM-N-7b (AGM-12C) Bullpup B with a 1,000-pound warhead. It was 13.6 feet long and weighed 1,785 pounds. It had a range of up to nine miles. To avoid the possibility of airframe damage from its more powerful rocket motor, the bigger Bullpup, like the Tiny Tim, was dropped from an ejector rack and fired when a lanyard attached to the rack reached full extension.

Iron Hand and the Shrike Missile

The effectiveness of Soviet surface-to-air missiles encountered in Vietnam generated countermeasures besides simply jamming the search and track radars. The most gratifying and practical was the radar-seeking missile, which would home in those radars. (Another much more challenging solution that would soon be pursued was the reduction of radar cross-section to the degree that the aircraft was undetectable, at least from certain aspects at certain frequencies.) The first U.S. anti-radiation missile was quickly developed from a Sparrow air-to-air, substituting a new seeker head. The resulting AGM-45A Shrike only weighed about 400 pounds. When it detected radar, it transmitted a tone of the pulse repetition frequency of the radar to the pilot's earphones and provided a bearing to the radar via his ADI. It had a relatively small warhead, but it was enough to shred a radar antenna, temporarily eliminating the enemy's ability to control a SAM.

The best available airplane for the mission, code named *Iron Hand,* was determined to be the A-4 Skyhawk, with the first Shrike-armed aircraft deploying in 1965. The *Iron Hand* A-4s would go out ahead of the strike itself, trying to attract the attention of the North Vietnamese SAM sites. When illuminated by radar, the A-4 pilot would launch a Shrike toward it. If the radar operator didn't shut his radar off, the missile would probably home in on the antenna and obliterate it. Collateral damage was light, however, and the antenna easy to replace. This was eventually rectified by adding a smoke marker that went off when the Shrike hit, allowing an immediate follow-up strike against the radar site.

The early Shrike had limitations beyond its small warhead. The A-4 pilot had to aim and launch the Shrike almost directly at the radar emission because the seeker head was not gimbaled. A significant drawback was limited range, which sometimes resulted in the radar site and the A-4 pilot having launched missiles at each other at the same time. However, if only one SAM was involved, the launch site would usually shut down the radar, forcing a draw, since neither the SAM nor the Shrike could continue to guide. The missile could also only home on a fixed frequency range, which might not match that of the radar site needing to be engaged. The range limitation was addressed in a subsequent modification introduced in the early 1970s, which incorporated a boost/sustain rocket motor that increased the missile's range to 25 miles.

Walleye

In a reversion back to unpowered, television-guided Glombs, NOTS began development of the Walleye in the early 1960s. Like the World War II weapons, the guidance was by video, only instead of a TBM or PV filled with avionics and an additional crewman to do the guidance, it was provided by the missile itself after being dropped from a single-seat airplane. The attack pilot set up the target picture on his multi-mode APQ-116 radar display and then transferred control to the missile's guidance system. When dropped, the Walleye would then steer itself to the selected image.

At 1,140 pounds, it was a heavy weapon with an 825-pound warhead and required a stores station to carry the data-link pod, which weighed 600 pounds. Because of its Circular Error Probable (CEP) of 10 to 20 feet, however, it was easily worth three conventional 500-

The Bullpup was thought to be the answer to both being able to stay out of range of antiaircraft fire and increasing accuracy. After launching the missile, the pilot guided it to the target by manipulating a small control stick to keep a flare in the tail of the missile superimposed over the target. (U.S. Navy via Hal Andrews collection)

pound bombs, not to mention the vulnerability reduction of dropping it at a distance from the target and not having to follow it in to guide it. It could only be used against targets, however, that could be discerned and tracked by the guidance system. This meant a high-contrast target that was well-illuminated and not obscured by smoke or dust. The range depended on the drop altitude and the ability to clearly see the target on the TV, but was on the order of a few to 15 miles.

The first combat use of Walleye occurred on 11 March 1967, when A-4Es of VA-212 from *Bon Homme Richard* successfully attacked military barracks and small bridges. The following day the target was the infamous Thanh Hoa Bridge. However, three direct hits only resulted in superficial damage, demonstrating its limitations. Against relatively soft targets like power plants, it was the weapon of choice in daylight.

In 1972, the punch of the Walleye was significantly improved. The Walleye II weighed twice as much as the Walleye I, but it also incorporated a 1,900-pound warhead. In October, it was used to destroy what was left standing of the Thanh Hoa Bridge.

A subsequent improvement to the Walleye allowed the pilot to drop it first and designate the target while it was en route. The designation could also be accomplished by another pilot in an aircraft carrying a data-link pod. This version was known as the Extended Range Data Link (ERDL). The Walleye II ERDL also had slightly larger wings than the standard Walleye, which gave it a maximum range when dropped from a high altitude of more than 30 miles.

Chapter Seven

SUPERSONIC STRIKE

The Enterprise deployed in August 1962 with some of the newest airplanes in the Navy, including the first full cruises of the F4H-1 and A3J-1. Also aboard in this picture were the latest F8U, the -2NE, and the Kaman HU2K, the turbine-powered helicopter, which was replacing the venerable HUP for plane guard. However, the Navy still could not do without the payload and endurance of the AD-6. (Vought Archives)

In the mid 1950s the Navy began programs that would lead to its first supersonic attack aircraft. The USAF was already in the process of equipping squadrons with the supersonic F-100C as a tactical fighter-bomber in 1955. It was followed by the F-100D, which was even more capable in the strike role. In 1952, the USAF had contracted with Republic Aircraft Corporation for the F-105. It was intended from the beginning to be a tactical nuclear bomber, carrying the store in an internal bomb bay for maximum speed capability. Although designated a fighter by the Air Force, in the Navy it would have been considered to be an attack aircraft. It flew for the first time in 1955 and was being operated in squadron strength in 1959. Since it was unencumbered by the need to take off and land from an aircraft carrier, it had a relatively small wing that allowed for a relatively smooth ride at high speeds and low altitudes. Its employment

The USAF F-105 was designated as a fighter, but all models like the YF-105B pictured here were actually designed as attack aircraft with a bomb bay for internal carriage of a nuclear weapon. (National Archives)

assumed the availability of a runway at least a mile long, however. Powered by a single Pratt & Whitney J75 engine, it was capable of Mach 1 at low altitudes and Mach 2 at 35,000 feet, albeit with no external stores.[1]

At the time, supersonic penetration of enemy defenses was increasingly considered by some to be necessary for survivability, in Congress if not in combat. According to George Spangenberg, Roy Eisner "a very emphatic strong-willed guy" in OpNav, "was con-

vinced that unless we had a supersonic delivery capability that naval aviation would disappear."[2] A supersonic dash capability greatly complicated interception by defending fighters or the early surface-to-air missiles. As or even more important, it matched the performance of the Air Force bombers competing for budget in Congress. The Navy was only just deploying jet attack aircraft, and these were only supersonic in a dive. Fortunately, it had almost completed the process of developing the changes required to aircraft carriers—the

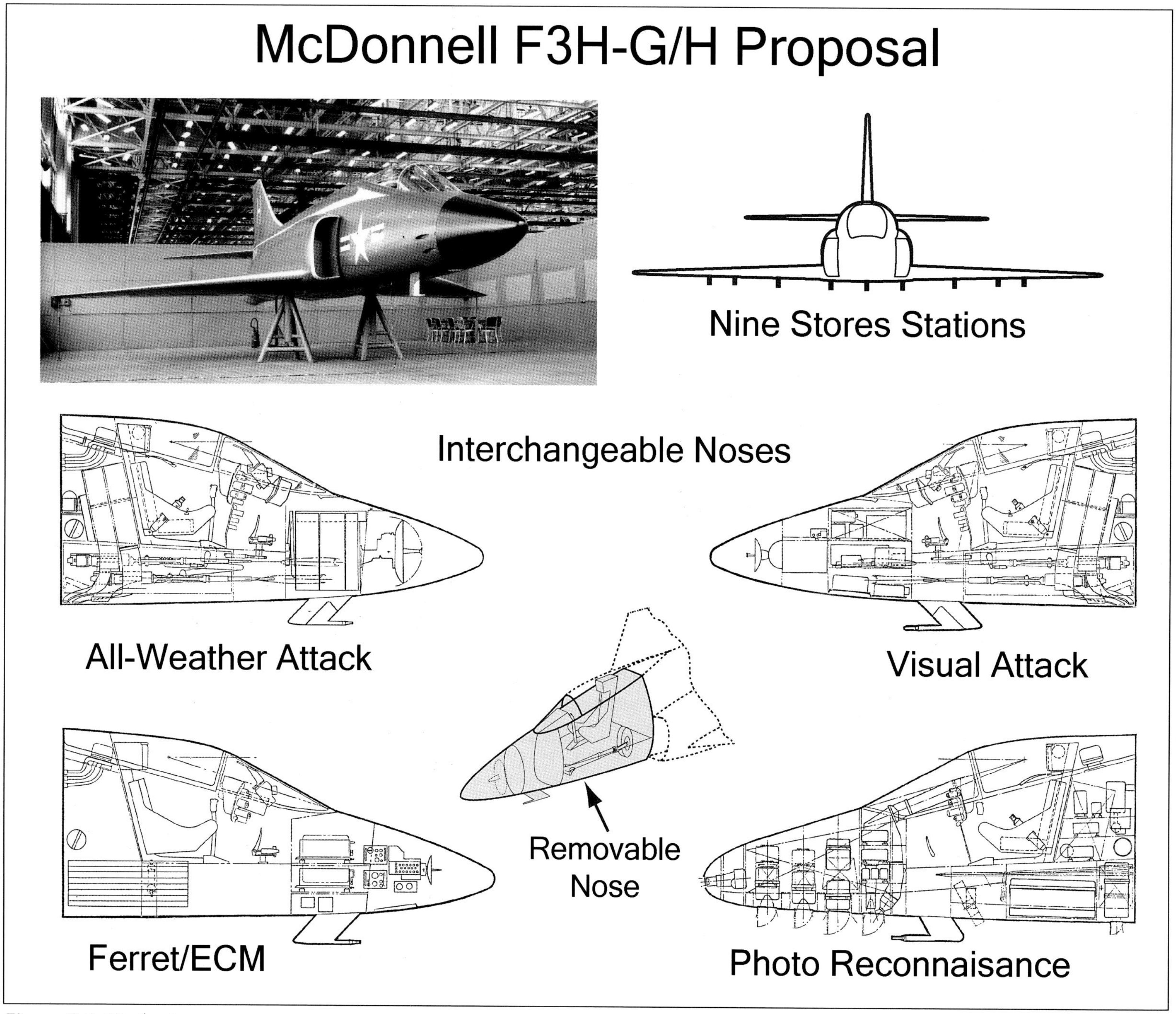

McDonnell F3H-G/H Proposal

Nine Stores Stations

Interchangeable Noses

All-Weather Attack

Visual Attack

Removable Nose

Ferret/ECM

Photo Reconnaisance

Figure 7-1. (Author)

steam catapult, angled deck, and mirror landing system—to launch and recover supersonic airplanes, necessitated by the need to defend the fleet and strike airplanes with fighters of comparable performance to their land-based adversaries.

As a result of the OpNav position, in 1954 the Attack Branch in the Bureau of Aeronautics began evaluating informal proposals for supersonic attack airplanes.

McDonnell AH

After McDonnell lost the supersonic day fighter competition that resulted in the F8U Crusader, it embarked on an aggressive marketing campaign to sell the Navy something else. Production of its F2H Banshee was ending and that of the F3H was just beginning, with its future not at all certain due to problems with its Westinghouse J40 engine. In September 1953, McDonnell provided a supersonic, all-weather, general-purpose fighter study to the Navy, featuring both single and twin-engine airplanes powered by a variety of afterburning engines, all iterations of the basic F3H configuration. The two that they spent the most effort on were powered by two engines like the F2H. The F3H-G was powered by the new J65 and the otherwise identical F3H-H by the even newer, still-in-development J79. One of the features of this design was an interchangeable nose, including the cockpit, allowing an aircraft to be reconfigured overnight from a fighter to an attack airplane or vice versa. The basic airplane had nine stores stations for maximum weapons capability. Each nose contained a radar, sight, and other equipment depending on a particular mission, such as four 20mm cannons, 54 2-inch folding fin rockets, or an ejector for photoflash cartridges. In addition to the noses shown, McDonnell included a single-seat all-weather interceptor alternative and a two-seat configuration.

The interchangeable nose was McDonnell's answer to the increasing number of dedicated mission types in a carrier's air group in the early 1950s that required a corresponding decrease in the number of fighter and attack airplanes aboard. A typical carrier air group deployment might now include detachments of aircraft for plane guard (helicopters), airborne early warning, electronic countermeasures, all-weather attack/antisubmarine warfare, all-weather air defense, photoreconnaissance, Regulus control, nuclear strike, etc. Using the interchangeable nose concept, a carrier could deploy with a slightly smaller air group and various extra noses. The task force commander could maximize or reconstitute his combat air patrol, interdiction, reconnaissance, etc., capability as required. For example, if one of the reconnaissance jets was lost or damaged beyond repair, another could be created using a spare reconnaissance nose to replace it.

In 1954, BuAer initiated a program to develop a single-seat, all weather, supersonic attack-fighter to replace both the Vought A2U for attack and the McDonnell F3H-2 as an all-weather fighter. It evaluated proposals in hand of the F3H-G/H, a single-engine F3H derivative, the Grumman 98D (a variant of the Tiger then designated F9F-9), and a marinized North American F-100. The F3H-G/H design study won this informal competition. McDonnell was provided a letter of intent in September 1954 for the AH-1. While BuAer and McDonnell were ironing out specification details like whether the J65 or J79 engine was to be used, BuAer and OpNav were still debating the mission requirement, which had apparently never been established. BuAer wanted to develop a general-purpose fighter while OpNav was coming to the conclusion that there was an urgent requirement for a high-performance fleet defense fighter armed only with missiles. As a result, the day before McDonnell signed the AH-1 specification on 27 May 1955, the Deputy Chief of Naval Operations (Air) signed out a letter to BuAer rejecting its 15 April recommendation that the AH-1 be armed with "guns, rockets, missiles and special weapons with a future capability for the 'ding dong' atomic rocket." OpNav requested that BuAer redesignate the AH-1 as the F4H-1 and stipulated that it be armed only with air-to-air missiles, although "a special weapons capability is also desirable provided it can be obtained

without imposing a significant penalty on fighter operations." Nevertheless, the F4H eventually became a general-purpose fighter as BuAer had intended.

NAGPAW Becomes the A3J

In 1954 North American was promoting its equally innovative North American General Purpose Attack Weapon (NAGPAW) concept. The initial proposal was single-seat and optimized for a low-altitude, high-subsonic speed profile. It incorporated three innovations—inertial navigation, rearward ejection of a bomb, and an auxiliary rocket engine for the getaway. Inertial navigation used a combination of a known starting point, motion sensors, and a computer to determine the airplane's current position. A low-resolution radar provided for a midpoint update of the inertial navigator for increased accuracy. The most unusual feature of the NAGPAW was the weapon release. It was stored internally for minimum drag, but instead of being dropped, it was fired rearward out of the airplane's fuselage. The opening was originally closed off with a two-piece clamshell door. According to the North American patent application, Weapon Delivery Method and Means, filed 1 December 1955 and patented 4 April 1961:

> With special stores...the conventional bomb bay or internal storage method is by far the most effective and practical. Nevertheless, utilization of such a bay having internally retractable bomb bay doors located substantially centrally of the fuselage has certain disadvantages. Thus, among other things, there is required elaborate sway bracing and suspension systems, fall clearance space, internally retractable doors and a

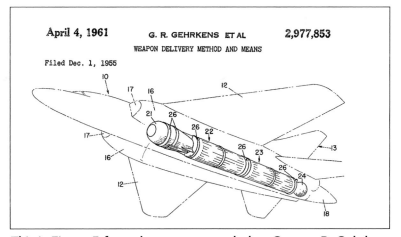

This is Figure 5 from the patent awarded to George R. Gehrkens and Frank G. Compton, which depicted the major elements of the longitudinal bomb bay. Item 18 was the "generally clam-shaped bomb bay doors"; 19, the armament tunnel; 21, the weapon; 22 and 23, the fuel tanks; 26, "plastic bands or the like" to brace the elements of the stores train; and 24, a countermeasure device. (Author's collection)

The A3J mockup had a few features that were changed. Those obvious here are the twin vertical fins, the clear canopy over the B/N, the curved outboard side of the engine inlet, and the extended windscreen. Two tails apparently provided hangar height clearance without folding. When the North American engineers refined their weight estimate, they probably found that a single folding tail was lighter than two non-folding tails. (Terry Panopalis collection)

close-out platform, each of which contributes measurably to an increase in airplane size and weight. Further, with the ever-increasing use of airplanes traveling at supersonic speeds, and frequently being required to release the internal stored weapon at critical attitudes, there arises the problem of guaranteed weapon separation. In addition, it has been difficult in the release of special stores from conventional bomb bays to avoid imparting an undesirably very high initial load on the weapon with the relative short ejection stroke permitted.

The attack branch was interested, but it wanted an airplane that was supersonic and also capable of being launched with zero wind over deck if required to initiate a nuclear strike on short notice even if the carrier was at anchor. It awarded North American a study contract in January 1955 to update its NAGPAW design with those requirements.

In retrospect, it seems like the BuAer Attack Branch replicated the McDonnell design that the Fighter branch had appropriated, except for the bigger wing for zero WOD launch and the internal bomb bay. The need for speed resulted in the selection of two afterburning J79-GE-8 engines, the same as the F4H. In addition, a rethink of the mission workload resulted in the addition of a second crewman. Both the A3J and the F4H therefore featured variable geometry intakes, boundary layer control over the flaps for slower approach speeds, tandem crew stations for minimum drag, and advanced mission avionics. Both airplanes were capable of Mach 2 at altitude and being supersonic at sea level. As it turned out, the F4H and A3J were developed in parallel, with major program milestones only separated by a few months, culminating with a joint first deployment. Although

far more F4Hs were built and garnered more world records, the A3J was an equally advanced design and far better looking.

	F4H*	**A3J-1****
Payload	Four Sparrows	One Mk 28
Weight (lb)	580	1,885
Internal Fuel (lb)	13,178	19,074
Gross Weight (lb)	43,072	55,160 Internal Fuel
Maximum GW (lb)	49,311	56,293
Wing Area (sq ft)	530	700
Combat Radius (nm)	410	685 Subsonic Cruise
Maximum Speed (kts)	1,220@36,000 ft	1,147@40,000 ft

* SAC dated 30 April 1960
** SAC dated 15 April 1961

In addition to the standard (for supersonic aircraft) stabilators, the A3J also featured an all-moving vertical fin for yaw control. In lieu of ailerons, roll control was accomplished with spoilers in conjunction with slots in the wing and "deflectors" that guided airflow into the slots as shown in Figure 7-2. It also had an early version of fly-by-wire control system, albeit with a mechanical backup. Of the innovative flight control system features, however, only fly-by-wire proved to be worth emulating.

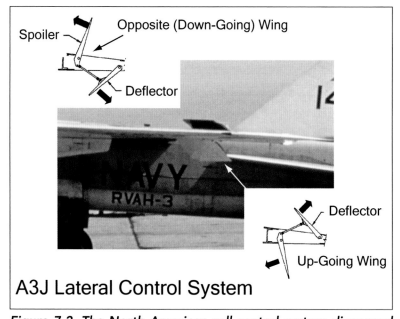

A3J Lateral Control System

Figure 7-2. The North American roll-control system dispensed with conventional ailerons, which caused wing twisting at high speeds, and added an extra feature to spoilers to provide for a roll axis coincident with the aircraft centerline. In essence, some air on one wing was diverted from the upper surface through the wing to the lower surface to increase its lift, and vice versa for the other wing to reduce its lift. (Author)

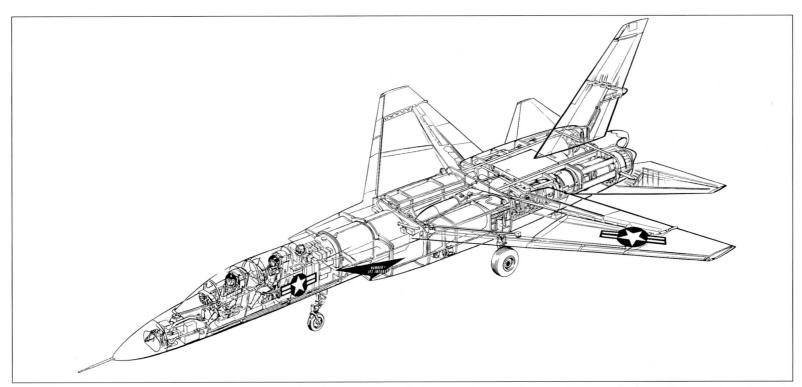

The A3J was large, but there was little unused volume, given all the avionics, mission equipment, fuel, engines, and an internal bomb bay. (U.S. Navy via author's collection)

The stores train was ejected rearward at about 50 feet per second to ensure positive separation. The hardware falling at the far right is the tunnel cap, which is attached to an electronic device that would jam or misdirect air-defense systems. (Terry Panopalis)

A typical A3J-1 stores train. On the left are the stabilizing fins that deployed upon ejection. On the right is the nuclear device, in this case an Mk 27 shape. In between are two fuel tanks. (Jay Miller collection)

A television camera was mounted in a clear dome under the nose of the A3J. It substituted for the periscope provided on the AJ and the A3D, but in this case the view was available to both the pilot and the bombardier/navigator. (Jay Miller collection)

When the large trailing-edge flaps were lowered for takeoff and landing, engine bleed air was blown over them for boundary layer control, a new aerodynamic concept in the mid-1950s and only practical with powerful jet engines like the J79. The leading edge flaps were set at 0 degree deflection when supersonic mode was selected on the flap control and drooped five degrees when it was set at cruise. Yet another setting provided for lowest fuel consumption for holding. With the trailing-edge flaps down, the inboard and intermediate leading-edge sections deflected to 30 degrees down and the outboard to 20 degrees.

For weight and balance purposes, the heavy bomb needed to be carried near the aircraft's desired center of gravity. The space in the tunnel behind it was utilized for fuel tanks, which would normally be carried as external stores. The fuel in these tanks would be used first, and then they would be jettisoned along with the bomb as an assembly. This was referred to as a stores train and included a set of extendable fins on the back of the most rearward fuel tank. After the tail cone was jettisoned, an ejection gun fired the stores train rearward at 50 feet per second, or about 30 knots, which provided for positive separation of the store and a short duration of the center of gravity shift aft as the bomb departed. The innovative concept was also compatible with both the loft toss and over-the-shoulder deliveries.

One stores option was an early version of a spoofing device that would be loaded at the aft end of the tunnel and ejected just before the bomb itself. Hopefully, the radar-guided enemy air defenses would focus on it instead of the A3J now prominently high in the sky. Another option for the space in the tunnel aft of the stores train was a rocket motor. This would provide climb performance and acceleration so that the A3J came over the target higher and faster than any interceptor could engage it. The oxidizer would be carried in the stores train tanks. When the target was reached, the rocket would be jettisoned, followed by the stores train ejection.

The bombing/navigation system was the AN/ASB-12. It included an inertial autonavigator, general purpose radar, a closed circuit tele-

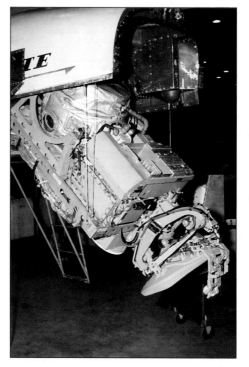

The A3J's radar antenna and a portion of the electronics associated with the AN/ASB-12 bombing/navigation system were readily accessible to troubleshoot problems. Unfortunately this was all too necessary due to the low reliability of the system. (Jay Miller collection)

vision (in lieu of a periscope), and a "versatile digital analyzer," VERDAN. The system provided navigation independent of outside aids and for weapons, computed trajectories, and provided an automatic release. A terrain avoidance mode was provided. The pilot was also provided with one of the first heads-up displays, called a Pilot's Projected Display Indicator.

Since the A3J would not be escorted, self-protection received emphasis. One unusual feature was the use of IR detectors mounted on the wings and fuselage. One was located on each side of the fuselage in the vicinity of the cockpit, and two more were mounted at the wing trailing edge in the vicinity of the wing folds, facing aft. During

This FJ-4 has been modified to add a rocket engine for evaluation of thrust management and other issues associated with its prospective application to the A3J-1 mission. (U.S. Navy via Hal Andrews collection)

BIS trials, they weren't found to be effective in detecting approaching fighters, even those in afterburner; the fuselage detectors provided excessive false warnings due to the sun. The APR-18, a detector of fighter-type radar emissions, was somewhat more effective. Self-protection also included the ALQ-41 and ALQ-51 jammers for X-band and S-band radars, respectively.

The A3J mockup review was accomplished in March 1956. The rocket motor option was canceled in early 1958 after a review of the impact on range and carrier suitability versus the benefit in avoidance of the bomb blast. According to George Spangenberg: "I had written a memo after getting some performance checked and we figured that with the radius of the airplane we were going to be lucky if we got to the outer ring of destroyers. The super performance rocket just took up too much fuel."[3]

The Vigilante was also intended to carry the XASM-N-8 Corvus missile, a relatively long-range, nuclear-armed weapon that would either home on an enemy radar emission or be guided to a target illuminated by the radar in the launch aircraft. It was to have a range of 170 nm as an anti-radar missile and 100 nm as a guided missile. The warhead was a W-40 with a 10-knot yield. Powered by a rocket motor, the missile weighed about 1,750 pounds. The contract was awarded to Temco in January 1957. First launch, from an A4D Skyhawk, occurred in July 1959 with fully guided flights by March 1960, but the program was canceled in July when the responsibility for all long-range nuclear air-to-surface missiles was taken over by the Air Force.

The A3J's first flight was made on 31 August 1958 with the airplane being taken supersonic less than a week later. Flight test was marred by the loss of the second aircraft in June 1959 due to a hydraulic and electric systems failure, but the North American pilot ejected safely. Another, BuNo 147851, crashed on 11 January 1961 at Patuxent River during the checkout of Maj. George Bacas by Lt. William Fitzpatrick who was riding along in the rear seat. Both were killed when Bacas crashed while attempting to land following a flap-control system failure. In March, LCdr. William Grimes apparently had a failure of both flight-control hydraulic pumps in BuNo 146700 while in the landing pattern at Patuxent River. When he attempted to eject, the unoccupied rear seat functioned properly, but his seat failed to fire due to an assembly error. He died in the crash. There was

The ASM-N-8 Corvus was a long-range, radar-homing, air-to-surface missile with a liquid-propellant rocket engine. It was to be armed with a small nuclear warhead with a 10 kT yield and have a gross weight of approximately 1,750 pounds. Temco received a development contract in early 1958. The first launch was from an A4D in July 1959 at Point Mugu. However, the program was canceled in 1960 after program responsibility was transferred to the U.S. Air Force in accordance with a realignment of roles and responsibilities. (Vought Heritage Center)

Although it was broad of beam because of the two J79s separated by the stores tunnel, the A3J was sleek when viewed from the side. The compactness dictated by carrier basing resulted in large vertical and horizontal tails being mounted well forward. (Jay Miller collection)

another fatal crash in August at Kirkland AFB, New Mexico, when BuNo 147855 caught fire on takeoff and rolled out of control. Neither LCdr. Guggenbiller or Lt. Biehl were able to eject in time.

Initial at-sea carrier trials occurred in July 1960 on *Saratoga*. Subsequent testing was accomplished aboard *Forrestal* (November 1960), *Midway* (December 1960), and *Ranger* (February 1961), with BIS trials being accomplished back aboard *Saratoga* in September 1961. Launch with a C-7 catapult at full mission gross weight required

about 10 knots WOD—higher than desired, but a few knots less than the A3D-2 (CLE), a significant achievement. Approach speeds ranged from 127 knots at a gross weight of 36,000 pounds to 139 knots at the maximum weight, 42,000 pounds.

The F4H set most of the world altitude, time to climb, and speed records, but in December 1958 the A3J got to make one—an altitude of 91,450 feet while carrying a 1,000-kg payload, breaking the existing record by 27,000 feet. (The Phantom altitude record of 98,557 feet was set with no payload; it would have had to be carried externally, limiting the top speed and therefore the altitude that could be reached with the ballistic trajectory required.)

Service acceptance trials commenced in February 1960 and ended December 1962 with several airplanes being used. The BIS report dated 22 April 1963 compared guarantees versus results with the -2 engines installed (the -8 had more thrust):

Item	Guarantee	Results
Maximum Airspeed		
(at 5,000 ft)	608 kt	596 kt (-2.0%)
Longitudinal Acceleration	2.979 ft/sec^2	1.71 ft/sec^2 (-43%)
Specific Range	0.122 nm/lb fuel	0.1213 nm/lb fuel (-0.6%)
Weight Empty		
(at 40,000 ft)	32,671 lb	32,879 lb (-0.4%)
Supersonic Combat Ceiling	51,500 ft	53,200 ft (+3.3%)
Maximum Speed		
(at 35,000 ft)	Mach 2.04	Mach 2.04
Stall Speed	109 kt	108.5 kt (+0.5%)

Nuclear Weapons Trials

Nuclear weapons trials were conducted by the Naval Weapons Evaluation Facility (NWEF) at Kirtland Air Force Base, Albuquerque, New Mexico, from May 1962 through December 1962.[4] Qualification was accomplished of the Mk 27 (internal), Mk 28 (internal and external), Mk 43 (external), and Mk 57 (external) stores. The planned qualification of internal carriage of the Mk 43 store, the replacement for the Mk 28, was canceled in October 1962, probably due to the decision to redirect the program away from the nuclear weapon delivery mission.

Thirty-eight successful full-scale nuclear store and store train deliveries were made along with numerous ones with small practice bombs. Both internal and external drops were accomplished using loft (about 65 degrees) and over-the-shoulder (OTS) methods with a run-in of 580 to 600 knots at an altitude of 100- and 200-ft above the ground. One lay-down delivery was made at 580 knots and 850 feet above the ground. High-altitude supersonic internal and external releases were made at altitudes from 35,000 to 61,000 feet and airspeeds from Mach 1.4 to 1.8. With the exception of the OTS delivery, the A-5A was able to egress quickly enough to stay outside the thermal limit envelope for the standard yields and burst height. OTS deliveries required limitations.

First sea trials of the A3J-1 were conducted aboard Saratoga in July 1960. The device that appears to be hanging from the pitot of the A3J is actually the plane-guard helicopter, a Piasecki HUP, its pilot flying sideward off the starboard side of the ship to monitor the launch and be in position to pick up the pilot in the event that it is unsuccessful. (Terry Panopalis collection)

The major deficiency in the weapon delivery trials at NWEF was the accuracy, reliability, and operational suitability of the Bomb Director Set, not the linear bomb bay concept. Although NWEF documented several minor problems with linear bomb bay hardware and function, its report does not single the concept out as a problem in stores delivery versus that of the external pylon deliveries, in spite of the fact that there was a notable difference in delivery accuracy in the loft and OTS modes. The flight crew noted a pronounced forward acceleration when an internal store was ejected but no significant pitch variation due either to the acceleration or the momentary shift in center of gravity.

The ballistics of both the Mk 27 and Mk 28 stores were considered adequate following low-altitude internal deliveries to meet accuracy requirements so misses were attributable to the airplane's bomb-delivery electronics. Some high-altitude, high-supersonic ejections resulted in marginal stability, but weapon limitations were not exceeded except in

The big speed brake shown here on A3J-1 BuNo 145158 was deactivated on the prototypes and deleted from production aircraft. Its function was replaced by the simultaneous activation of the wing spoiler deflectors. This reduced weight and complexity, of which the A3J still had plenty. The nose of this airplane is filled with flight-test instrumentation instead of the radar and mission avionics. The aft cockpit is also occupied by flight-test hardware. (Jay Miller collection)

The pilot of this NATC A3J has landed almost exactly on centerline, probably during one of the first landings during this trials period because most carrier suitability testing is devoted to evaluation of less-than-perfect arrivals. The vapor is from the fuel-vent system. (Jay Miller collection)

one case when the fins failed to deploy on an Mk 27 stores train ejected at 61,000 feet and Mach 1.6, resulting in the breakup of the assembly.

The summary score card, compared to the guaranteed Circular Error Probable (CEPs), was as follows:

	External Store		Internal Store		
	Hit	Miss	Hit	Miss	
Loft	6	0	1	5	Internal Misses 2X CEP Guarantee
OTS	3	4	1	5	Internal Misses 2X CEP Guarantee
Laydown	0	1	0	0	Visual Low Altitude
Level	2	3	2	6	Radar High Altitude
Total	11	8	4	16	

In general, guaranteed CEPs were 1,000 to 1,500 feet, depending on the delivery specifics.

It was clear that internal stores' loft and OTS results were significantly worse than those for external stores, but no explanation for the difference was provided other than a brief discussion of range performance versus tables:

Range data reduced to no wind conditions were used to obtain no wind bomb range for each internal release. Results were compared with data available in published range tables. Comparisons were not definitive, in part because of limited data available in the appropriate range tables. In general the Mk 27 internal store train consistently fell short of published values. Mk 28 stores trains were less consistent and errors were of smaller magnitude. All ballistic data has been forwarded to NWL Dahlgren for detailed analysis.

Over-the-shoulder (OTS) delivery accuracy with a release at 120 degrees was adversely affected for both delivery modes by the degradation in airplane control response and effectiveness as the pitch exceeded 75 degrees. However, the external OTS misses were close and tightly grouped, with one exception attributable to a poorly flown profile. The internal misses were notably long; one by a mile, with a lot more scatter than could simply be handled by a ballistics correction.

The accuracy at high altitude for both external and internal releases was in part limited by the radar's ability to acquire the target reflector at altitudes above 40,000 feet and the inability of the bombardier to verify the target on the television sight. Again contrary to anecdotes of the store drafting after the Vigilante following release, causing delivery inaccuracy, the average miss of the external deliveries was actually more than that of the internal, although the former was almost 1,000 feet more than a mile and the latter, 9/10 mile, nothing to crow about when the CEP guarantee was on the order of 3/10 mile.

The Final Armament Trials Report dated 26 November 1962 reinforced the problem with the bombing system:

Major deficiencies of the AN/ASB-12(XN-2) Bomb Director Set which still exist include: unsatisfactory bombing accuracy with the F10F Bombing Computer in the maneuvering modes; excessive drift rate of the inertial autonavigator; and inability of the general purpose radar to provide an

The A3J-2 featured a raised mid-fuselage to increase the internal fuel capacity. The fairing in the armpit between the fuselage and the wing covered the piping for the addition of boundary layer control to the leading-edge flaps. On the production aircraft, the piping was routed on top of the wing adjacent to the fuselage. (Jay Miller collection)

adequate radar presentation because of noise modulation and range distortion of targets.

Conversion to an Important Supporting Role

In order to provide more range and external stores options, BuAer contracted for the A3J-2. The major change was the addition of more and larger fuel tanks in the upper fuselage, giving the airplane a hunchback appearance. The increase in fuel capacity totaled about 460 gallons. The wing was modified to have larger flaps and boundary layer control air emanating from the hinge point of the wing leading edge flaps. Another stores station was also added to each wing.

By the time of the A3J-2's first flight in April 1962, however, the decision had been made to redirect the Vigilante program to another mission requirement. With the advent of the submarine-launching Polaris ballistic missile, the A3J-1 strategic nuclear weapons-delivery mission was redundant. However, the design was to be adapted to the tactical reconnaissance mission. The announcement of the contract in March 1962 noted, "The change in configuration will not affect the ability of the aircraft to perform its basic attack mission." The resulting A3J-3P also retained the fuel capacity and wing changes introduced with the A3J-2.

Reconnaissance, both photographic and electronic, was a very important antecedent to a successful strike mission against an infrastructure target. Survivability and effectiveness depended on having the best possible understanding of the target and its defenses for proper mission planning. (Post-strike pictures were also important to determine if the mission had been successful or a re-strike was required.) Up until now, the important reconnaissance mission had been accomplished by two different aircraft: a fighter that had been modified to carry cameras instead of guns and a bomber modified with antennas and avionics that could intercept, record, and analyze radiated electromagnetic energy. This provided an assessment of the enemy's defensive electronic order of battle for use in planning strike-missions. Knowing where emitters were located allowed the strike-mission planners to develop ingress and egress routes that would minimize the reliance on the bomber's self-protection Electronic Support

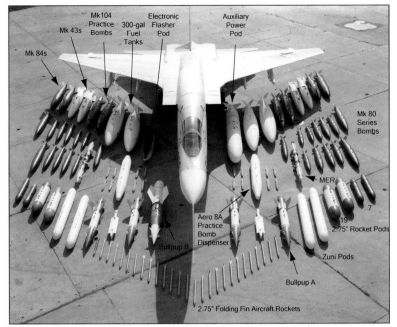

The A3J-2 was provided with an additional pylon under each wing for external stores, both conventional and nuclear. The auxiliary power pod was to be carried when the airplane might have to land at an airfield not equipped with the proper ground-support equipment. It contained a turbine engine that provided electricity, refrigeration for the avionics and the crew's pressure suits, and bleed air to start the J79 engines. (U.S. Navy via author's collection)

The RA-5C was intended to retain a nuclear weapon delivery capability, carrying the weapon on an external pylon, a capability qualified on the A3J-1. This Mk 28EX (for external carry) shape is being fit checked on the left inboard wing pylon since it was a slightly different configuration on the A3J-1. (Jay Miller collection)

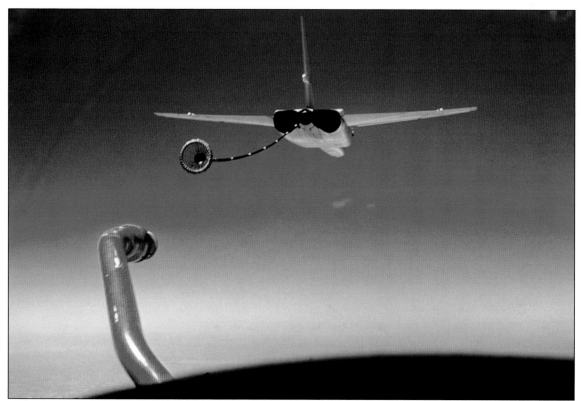

As it turned out, the A3J's innovative and unique stores tunnel was not used operationally for weapons. It did see limited use for inflight refueling, in this case of another A3J. (Robert L. Lawson collection)

The A3J-1 was also qualified for conventional weapons delivery, here with a fully loaded Triple Ejector Rack (TER) on the wing pylon. The chase plane is an F8U-2. (Robert L. Lawson collection)

Measures (ESM) equipment and properly position the electronic warfare aircraft that would be able to actively jam the enemy radars and communications.

The A3J-3P was big enough to carry an internal equipment pallet and a belly pod that could be equipped with cameras, side-looking radar, IR sensors, and electronic signal detectors. The ASB-12 attack capability was retained using the external wing stations for stores, although it was probably never used operationally. The first A3J-3P,

The Polaris missile met, if not exceeded, all expectations. Even if the A3J's troubled bombing system had worked as North American had envisioned, the likelihood of the carrier-based Navy retaining a strategic delivery mission was low. (U.S. Navy DF-SC-84-07332)

soon to be designated RA-5C, flew in June 1962. All the -2s were eventually completed and delivered: two as the A3J-2 (A-3B) for flight-test, four as YRA-5Cs without the reconnaissance pods to be used as interim trainers, and the remainder as RA-5Cs.

Although sleek to begin with, such as the A3J-1, at almost 80,000 pounds the RA-5C was second only to the A-3 in terms of maximum launch weight.

	A3J-1*	RA-5C**
Payload	One Mk 28	Recon Package
Internal Fuel	19,074	24,480
Gross Weight (lb)	55,160	65,589 Internal Fuel
Maximum GW (lb)	56,293	79,405
Wing Area (sq ft)	700	754
Combat Radius (nm)	685	475 Subsonic Cruise/ Internal Fuel
Maximum Speed (kt)	1,147@40,000 ft	1,120@40,000 ft

* SAC dated 15 April 1961
** SAC 00-110AA5-2

The RA-5C not only retained the anti-flash provisions in the cockpits for a time, but aircraft were also initially delivered, like this prototype, with pale blue "Navy" markings to reduce the thermal impact of a nuclear bomb explosion. Although gloss white undersides and control surfaces were standard, anti-flash markings were rarely applied to U.S. Navy aircraft. (Jay Miller collection)

The RA-5C was updated beginning with new deliveries in 1969. The modifications included higher thrust J79-GE-10 engines, enlarged inlet ducts, and extended wing root fillets. There were also detailed internal changes such as an improved landing gear and maintenance improvements.

A total 59 A3J-1s were produced, including the two prototypes, with 43 being rebuilt as RA-5Cs. All 18 A3J-2s eventually became RA-5Cs. These were followed by 79 new RA-5Cs, for a total of 156 aircraft. Almost half were to crash or be shot down.

George Spangenberg was very complimentary about the RA-5C's mission capability, if not North American's design prowess:

> (As a) reconnaissance airplane, (it had) reconnaissance capability orders of magnitude better than anything we had been able to do before. We always had photo versions, probably every one of the fighter airplanes...it was orders of magnitude better. Bigger cameras and electronic reconnaissance were incorporated as well. And in connection with that development they did that reconnaissance capability aboard ship. It became a separate unit unto itself. From an airplane's standpoint, the airplane had all kinds of troubles. Detailed little problems had been going on in North American for a long time. I was never sure exactly why. They appeared not to have enough designers that had made mistakes before. They were making them all for us. The AJ-1 was a prime example. They put the hydraulics on top, electronics on the bottom so the hydraulic fluid leaked all over the airplane instead of putting (the hydraulics) on the bottom where at least you could put a drain hole in and let (the hydraulic fluid) run outside.[5]

Operational Usage

In January 1960, the Deputy Chief of Operations for Air recommended that the A3J-1 be deployed only in the Atlantic Fleet. At that point, the first Polaris-carrying submarine had been commissioned and the total A3J-1 buy was expected to be only 77 airplanes, allowing an operational fleet of 48 to 52 airplanes, or four squadrons. "Since the A3J was designated as a nuclear delivery vehicle capable of penetrating heavy defenses, it is imperative that these few aircraft be utilized in the area of greatest threat of nuclear war (Europe) and against the hard core targets of the USSR and western satellites."[6]

The first squadron to receive the A3J-1 was VAH-3 in June 1961, even before BIS trials and qualification was complete. VAH-3 was the training squadron, responsible for checking out the pilots and bombardier/navigators in the Vigilante. The first deployable squadron was VAH-7, which was the next to receive aircraft. It departed Norfolk with 12 airplanes in August 1962 aboard *Enterprise,* the Navy's first nuclear-powered aircraft carrier, deploying along with the first F4H squadron to go to sea. It was somewhat premature since weapons-delivery trials had not yet been completed, not to mention correction of the lack of reliability of the AN/ASB-12 bombing/navigation system, along with limitations of its radar's functionality in cloud and rain. It

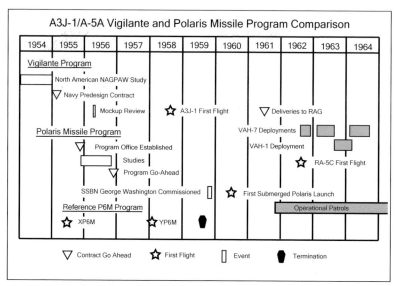

Summary Schedule—Strategic Programs. (Author)

The Enterprise *air group first used the tail code "AF" until, so the story goes, a visiting congressman said that he didn't know that Air Force airplanes operated from Navy carriers. The tail code was swiftly changed to "AE." The wing stores on 704 and 705 are Aero 8A practice bomb dispensers.* (U.S. Navy via author's collection)

doesn't appear that there was ever much confidence that the aircraft could be used to find and hit a target with an acceptable degree of accuracy even for a nuclear weapon. Before the A-5A, *née* A3J-1, was withdrawn from the nuclear strike mission, VAH-7 was to make only two more deployments, and VAH-1 only one. The only operational use of the linear bomb bay besides additional internal fuel capacity appears to have been to carry a buddy tanker pod for inflight refueling.[7]

A small Ryan reconnaissance drone was evaluated during a Ranger deployment. It was launched from a mobile zero-length trailer by JATO and recovered from the water after a parachute landing. (U.S. Navy via Craig Kaston)

The A3J-1 did not enjoy as good a reputation as the F4H for carrier suitability during its brief operational career. In spite of its bigger wing, it had a high approach speed and was a demanding airplane to land even on the big decks to which it was restricted. Ramp strikes with fatalities were not uncommon. CDR Bud Gear was the first. An experienced aviator, he had a total of 4,603 hours and more than 400 traps, of which 33 were at night. He had launched solo for night qualification from *Franklin D. Roosevelt* in January 1962 and hit the round down on his first approach. VAH-7 lost a plane and crew, Lt. Charles Kruse and Lt. (j.g.) Clarence Cottle, to an *Enterprise* ramp strike in February 1963, again at night, only two weeks into the second deployment.

RVAH-5, the successor to VAH-5, was the first squadron to deploy in the RA-5C when *Ranger* departed Oakland, California, in August 1964. As a result of the Gulf of Tonkin incident, it became the first RA-5C squadron to fly over Vietnam. By happenstance, *Ranger* also hosted the last Vigilante deployment, accomplished by RVAH-7 and concluded in September 1979, 14 years later.

In between, the RA-5C provided a carrier-based reconnaissance capability that was not available before and not equaled since, but at great cost. During the Vietnam War, eight RVAH squadrons made 32 combat deployments. Of the small fleet of airplanes, 18 were lost in combat and an additional five were destroyed in accidents. It was the highest loss rate of any Navy airplane involved in the war. Its seemingly high vulnerability was in part caused by the dictate that a strike be followed by an RA-5C flyby for a photographic strike assessment. The foreknowledge that an airplane would be coming by soon and the ability to focus all available anti-aircraft capability on that one airplane maximized the North Vietnamese's air defense chances of downing it.

In order to obtain the essential pre- and post-strike photographs without risking the loss of a crew and an expensive airplane, the Navy experimented with a small, Air Force-developed, unmanned reconnaissance drone, the Teledyne-Ryan AQM-34 with the Ryan designation 147SK. It was fired off the carrier from a zero-length launcher with a rocket assist. Guided to an initial checkpoint by an E-2 Hawkeye, from there it flew its mission and returned to a recovery point autonomously. It would then slow and deploy a parachute, to be recovered in midair by a helicopter or from the water. Recovery was necessary to access the film taken during the flight. There were about 30 operational flights over Vietnam between November 1969 and May 1970 flying from *Ranger*. The experiment was terminated with the end of *Ranger's* deployment. Even when the drone flew a profile precisely enough to take pictures of the desired objective, as opposed to some portion of the surrounding countryside, the quality was not as good as provided by the RA-5C. It also didn't have the other onboard surveillance systems, being a much smaller aircraft. Operationally, it was an unwelcome disruption to flight-deck operations to set up and launch. One went astray in February and wound up parachuting down on Hainan Island, Peoples Republic of China. For some reason the Chinese did not complain about the incident but it was another strike against the concept.

The F-4's Supporting Role

The F4H was redesignated the F-4 in 1962. Although its primary role was fleet air defense, like almost all of the Navy's fighters it could also be used as an attack aircraft. The original definition of this capability when the jet airplanes were introduced was General Purpose Fighter. Even though it lacked a gun and almost always carried a large external fuel tank on the centerline station, it was equipped with four wing pylons that could carry several different types of stores. During the Vietnam War, it was often used to augment the dedicated strike aircraft.

However, the air-to-ground role was not initially practiced in the F-4 community. The F-4's optical sight had even been removed in some squadrons for better over-the-nose visibility. When VF-96 was specifically assigned the secondary mission of flak suppression for its 1964 deployment, the squadron arranged at the last minute to have the A-1 Skyraider gunsight substituted for the F-4's simpler one to increase aim point adjustability. Subsequent deployments of F-4 squadrons were somewhat better prepared for air-to-ground missions. Late in the war, the F-4s were used to drop Air Force-developed laser-guided bombs, one Phantom being the bomber and another the designator, with the Radar Intercept Officer using a hand-held device.

Ironically, given all its world records for speed compared to the A3J's one for altitude, the F-4 was at a disadvantage operationally. In Vietnam, an F-4 would always escort the RA-5C on its reconnaissance and post-strike evaluation missions. In order to carry the necessary fuel for range and endurance and provide an air-to-air alternative to the Sparrow, the F-4 was burdened with the big external centerline tank and at least two Sidewinders. The RA-5C, however, was almost always clean, with its additional fuel carried in three large fuel tanks located in the linear bomb bay. As a result, with the North Vietnamese taking umbrage at having their picture taken, the RA-5C would depart the premises at a notably higher speed, leaving the escort behind. The Viggie pilot was also less concerned about the location of the nearest tanker. On the other hand, he would subsequently be more concerned about the upcoming carrier landing than his erstwhile escort in the F-4.

An RA-5C is ready for launch from John F. Kennedy. The pilot's heads-up display was eventually removed from the Vigilante, providing one of the most unobstructed forward views of any carrier-based aircraft. (Robert L. Lawson collection)

The F4H was initially intended to carry bombs as demonstrated by this early flight-test photo. However, the capability languished until the Vietnam War required that almost every aircraft in the air wing periodically drop its weight in bombs. (U.S. Navy via author's collection)

F-4 pilots quickly adapted to their supporting role in air-to-ground missions during the Vietnam war. This VF-21 F-4B from Midway is dropping Snakeyes that have the fins wired shut. (Terry Panopalis collection)

CHAPTER EIGHT

ALL-WEATHER ATTACK

After a somewhat rocky start in combat, the A-6 Intruder took over as the main battery of the U.S. carrier strike force. (U.S. Navy via author's collection)

Even after the introduction of jets for the attack mission, the AD Skyraider continued to not only be a highly valued close air support and all-weather attack airplane, but provide strike mission support with electronic reconnaissance, radar jamming, and inflight refueling. It was so versatile that two different airplanes, the Grumman A-6 Intruder and the Vought A-7 Corsair II, eventually replaced it.

The Wide Body Skyraider

The multi-place AD-4N all-weather attack Skyraiders were replaced by the multi-place AD-5N beginning in 1954.[1] What was to become the AD-5 was originally proposed in December 1949 as an ASW airplane that merged the hunter and killer requirements into one platform. It used the same basic wing, engine, landing gear, and other

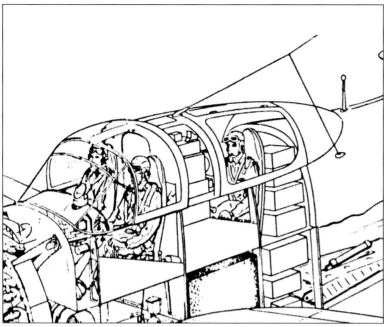

The AD-5 was a reconfiguration of the basic Skyraider to provide a crew compartment wide enough for side-by-side seating. This cabin was filled in different ways for various missions. The aft compartment of the utility version was big enough to accommodate four rearward-facing seats for transportation of personnel. (Author)

systems of the AD-4, but the fuselage was widened for side-by-side seating and a compartment provided aft of the cockpit to provide more interior volume and crew comfort than allowed by the cramped mid-fuselage of the basic Skyraider. The area of the vertical fin and rudder were increased by about 50 percent, in part to eliminate the need for the tail fins on the AEW version to compensate for the forward side area of the radome. The speed brakes on the sides of the fuselage were deleted, but the large belly-mounted speed brake was retained. The prototype, a conversion of an AD-4, first flew in August 1951.

Since the AD had already been adapted to a number of different multi-place mission requirements, the next step was the proliferation of the wide body derivatives of those missions. Douglas marketing brochures promoted the AD-5 as "History's first true 'multiple' or 'all-purpose' airplane." It was claimed "Conversions from one mission to another could be accomplished aboard Navy carriers in from two to 12 hours." That was true to an extent for a plain-vanilla day-bomber AD-5 and utility missions, but the AD-5N for night attack and the AD-5W for AEW came off the Douglas production line heavily equipped and hard wired for those particular missions. These configurations were not convertible to another except at a major overhaul facility.

The basic AD-5 was about 600 pounds heavier than the AD-4. The AD-5N configuration was 1,300 pounds heavier than the basic AD-5 because of the big AN/APS-31C search radar permanently attached under the right wing and other equipment. A combination

searchlight and sonobuoy/flare dispenser or a chaff dispenser could be carried on the left inner wing bomb rack depending on the mission. The APA-16 low-altitude bombing system provided automatic positioning of the search light on a selected radar target. ECM equipment included a radar/radio signal receiver that provided bearing to the transmitter and an APT-16 transmitter for jamming radar and radio transmissions. The crew consisted of a pilot in the left front seat, an ECM operator in the right front seat, and a radar operator in the "blue room" behind them.[2]

The missions assignable to the AD-5N included night attack, anti-submarine warfare, and radio countermeasures. The VC crews trained for all these missions, including all-weather delivery of a nuclear weapon. With water injection, a 1,660-pound store, and a 300-gallon drop tank, the takeoff gross weight was 23,205 pounds and the mission radius was 500 nm. Because of the gross weight increase without a corresponding increase in engine power, the idiot loop delivery was modified to be a wingover.

To more accurately reflect the all-weather mission, VC-33 and VC-35 were redesignated VA(AW)-33 and VA(AW)-35 in July 1956. As part of the transfer of all-weather attack mission responsibility to the air group squadrons, VA(AW)-35 became VA-122, the fleet replacement training squadron for the AD-6/7, in June 1959.

The increasing importance of the ECM mission and unique avionics required resulted in the conversion of 54 AD-5Ns into AD-5Qs. (VA(AW)-33 became VAW-33, flying the AD-5Q and providing detachments to air wings for ECM support.) The APS-31P installation was retained, but the dive brake was deleted to allow the installation of an APQ-33 ECM antenna. Other ECM antennas were added to the sides of the aft fuselage, and a tail warning antenna was added to the rear of the fuselage facing aft to detect emissions from fire-control radars. Mission pods were carried on the outboard wing pylons as required. There were positions for two ECM operators in the aft cabin and the controls, indicators, and analyzers spanned the width of the cabin and filled a center console, eliminating access between the cockpit and the cabin. The crewman to the right of the pilot was now the radar operator. The 20mm guns were originally retained and then removed along with the gunsight. They were reinstated during the Vietnam War to provide a modicum of self-protection.

The AD-5N designation was changed to A-1G in 1962, with the AD-5Q becoming the EA-1F. The Navy was to deploy relatively few AD-5s as day bombers, but they came in handy as a utility transport and for target towing. The electronic variants, on the other hand, became a fixture in air groups for many years and outlasted the single-seat ADs. The last Navy Skyraider missions were flown by VAQ-33 in EA-1Fs providing ECM support for the first *John F. Kennedy* (CV-67) deployment which ended in December 1968.

Grumman A2F Intruder

In October 1956 the CNO issued a requirement for a big subsonic, all-weather attack airplane to replace the AD-5N. In this case, there was a British-developed carrier-based airplane about to fly, the

Blackburn Advanced Naval Aircraft, which was very similar in capability and performance. In 1953, the Royal Navy had come to the conclusion that its new carrier-based nuclear strike aircraft, to be used primarily against ships at sea and ports, would have to approach and attack at very low altitude to avoid radar detection. The combination of compactness for carrier-basing, low-altitude cruise and attack, and long-range requirements virtually precluded afterburning engines and supersonic performance. However, unlike the later U.S. Navy development, the resulting Blackburn (which later merged with Hawker Siddeley) Buccaneer was otherwise optimized for near-sonic speed at sea level, featuring tandem cockpits, an internal bomb bay, area-ruling, retractable in-flight refueling probe, small wing, etc.

George Spangenberg remembered the program as being started to fulfill a Marine requirement for a very short takeoff from a land base. According to him:

> OSD approval for the program was held up until the Navy defined a secondary, long range attack mission and included the model in its plans for the future. At initiation, the design close support mission called for an endurance of one hour at sea level at a radius of 300 miles following a short take off (STO) with two 1,000-pound. stores, using partial fuel. For the Navy, the airplane had an estimated radius of (1,000) nautical miles with two external tanks and a single 2,000-pound. store, or a radius all at sea level, of 730 miles with four tanks. The short take off requirement of the Marines necessitated excellent low speed performance, while the Navy requirement demanded an efficient cruise arrangement; which combined to give the airplane its margin for growth and an unprecedently long production life.[3]

The takeoff and landing were to be accomplished over a 50-ft obstacle within 1,500 feet or less. The minimum acceptable maximum speed at sea level was 500 knots.

From a weapons standpoint, five pylons were to be provided, but no guns. The Operational Requirement dated 2 October 1956 included the statement "It is emphasized that the paramount feature of the weapons control system is that it have the capability of detection, lock-on, and guidance for attacking moving or stationary targets under non-visual conditions." Somewhat increasing the degree of difficulty was the stipulation to do this in a single pass with at least five mile accuracy, which was significantly better than a visual attack.

In an attempt to minimize the size of the aircraft, the Operational Requirement stated that:

> The aircraft shall be of such size and weight that it will operate without restriction from 27A class carriers and temporary airfields. In this development, every effort should be made to hold the full load weight to a maximum of 25,000 pounds.

Even if "full load" referred to the short-range, internal-fuel-only, close-air support mission, this was wishful thinking, perhaps engendered by the thought that it was a replacement for the AD-5N, which had a maximum gross weight of 25,000 pounds. As it happened, in production the *empty* weight of the successful design was 25,000 pounds.

Unlike the F4H and A3J programs, it was a formal competition, with Type Specification 149 being provided to virtually all the major military airplane contractors in the U.S. in March 1957. At that point, it hadn't been decided whether the new airplane would be single or twin engine or powered by turboprop or jet engines, so the maximum speed required was only 500 knots. Some manufacturers therefore responded in August with more than one design, although only one was a turboprop. Bell Aircraft even proposed a vertical takeoff and landing (VTOL) aircraft. Only the single-engine jets met the gross weight of 25,000 pounds, but they all fell short of more important requirements. By October, the short list contained only the twin-J52 engine powered proposals from Grumman, Douglas, and Vought, all traditional Navy contractors.

Douglas proposed a swept wing mounted at mid fuselage with the engine nacelles slung under the wing. The short field capability was achieved with water injection and JATO for takeoff and a parabrake for landing in addition to an installed engine thrust line angled down 10 degrees. The crew sat in tandem under separate canopies. The internal fuel capacity was 12,240 pounds.

Vought proposed a high-mounted straight wing with wing-mounted engine nacelles. The short field capability was enhanced by a drooped leading edge, full-span flaps, water injection, and retractable cascade-type thrust deflectors in the engine exhaust. The crew sat in tandem like the Douglas proposal, but under a single canopy. One unique feature was the wingtip pod mounting of the ECM antennas. The internal fuel capacity was 14,750 pounds.

Grumman's wing was swept and mounted at mid fuselage. However, unlike the Douglas and Vought designs, the engines were mounted adjacent to the fuselage. The short field requirement was achieved without water injection by the provision of tailpipes that tilted 23 degrees downward for takeoff and landing. To shorten the landing roll, it had nose-wheel braking. Unlike the tandem cockpit arrangement of the other two finalists, the crew sat side by side with the bombardier/navigator's seat a few inches lower and aft to minimize interference with the pilot's view to the right. The internal fuel capacity was 15,000 pounds. At a gross weight of 32,500 pounds for the basic design mission and 45,265 pounds for the long-range mission, it (as well as the other competitors) was significantly heavier than the AD-5N.

The BuAer technical assessment was that all of the designs on the short list met the maximum speed requirement, with the Grumman being six knots faster than the slowest, which was Vought, and four knots faster than the Douglas. In its estimation, none of the designs met the one-hour loiter on the short-range mission at the fuel quantity estimated by the manufacturer. None met the short takeoff distance of 1,500 feet over a 50-ft obstacle—Vought would have if its performance hadn't been penalized for loss of control at full thrust deflection if an engine failed. Grumman had the least shortfall of 181 feet, which would have been even less with water injection and Douglas the worst at 375 feet. Only Grumman met the landing distance, with Vought

The AD-5Qs were created by modifying existing AD-5Ns. These featured additional ECM capability as indicated by the rearward facing antenna under the rudder and a radome under the fuselage where the dive brake had been. (Robert L. Lawson collection)

The AD-5N was a very capable night-attack aircraft. The big APS-31 radar under the right wing provided for navigation and attacks on radar-significant targets or those with nearby radar-significant features. Most of the outboard stores pylons have been removed and covered with tape. (Steve Bradish collection)

The Blackburn Buccaneer was the Royal Navy equivalent of the A-6 Intruder in size and mission. It lacked the full terrain-following capability of the A-6 since its primary mission originally was attacking Russian surface ships at sea. (Vought Aircraft Heritage Foundation)

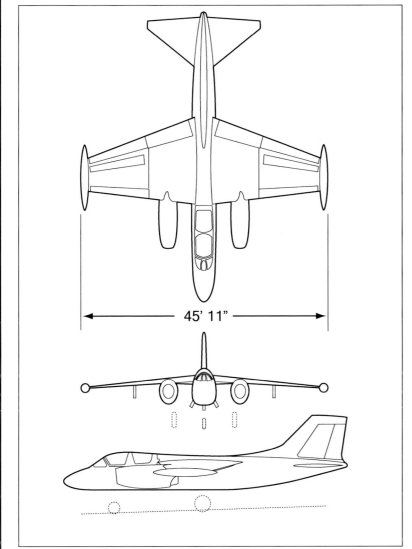

Vought's Model 416 faired somewhat better in the competition for a new Navy attack airplane with its innovative thrust deflector system, but its stores pylon arrangement was less well received. (Author)

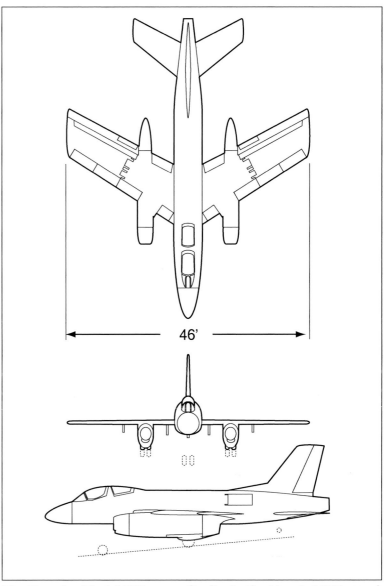

The Douglas Model 715 was a finalist in the Navy's 1957 attack competition but came in third. (Author)

again not being credited with full deflection capability and Douglas having to use the parabrake to achieve it. Only Vought met the 1,000-nm radius at the proposed fuel loading but Grumman's design exceeded it with four fuel tanks instead of the two desired. Douglas didn't meet it even with four tanks.

The summary conclusion was that Douglas came in third overall on performance, hurt by having the least internal fuel and no thrust deflection for takeoff and landing. The Douglas avionics suite was also judged to be more expensive and more of a new development, hence a higher schedule and technical risk. Grumman was rated the highest technically, with the least risky avionics proposal. BuAer evaluators also liked the side-by-side cockpit arrangement and lower

infrared detectability of the engine installation. The Grumman stores pylon arrangement was considered superior to Vought's since no pylon was located beyond the wing fold.

On 30 December 1957, Grumman was declared the winner of the competition with its design designated the A2F. The initial contract for design effort and a mockup was issued in February 1958. The detail specification wasn't signed until May 1958, but a delay while the final specifications were hammered out between the contractor and the Navy was not unusual.

Unlike the Buccaneer, the A2F had side-by-side seating for the pilot and bombardier/navigator, no bomb bay, no area-ruling, a fixed in-flight refueling probe, and a relatively high aspect ratio wing.

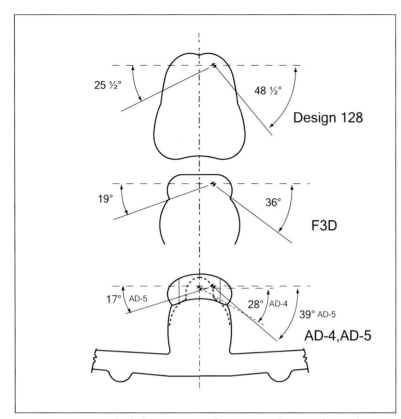

The Grumman proposal for the AD-5 replacement was only slightly different from the airplane that resulted. The refueling probe was relocated to be on the centerline of the aircraft in front of the windscreen and the wing leading and trailing edges were straightened, among other things. (Grumman History Center)

Grumman provided the Navy evaluators with a study to demonstrate that the proposed side-by-side seating arrangement provided adequate visibility to the pilot. The evaluators agreed and more importantly, the flight crews came to think of it as a superior arrangement for the two-man crew attack mission. (Author, from Grumman data)

The A2F-1 mockup review resulted in no major changes to the basic configuration. The size of the rudder and the drag brake location shown here proved to be unacceptable in flight test, however. The stores loaded from left to right are a Zuni pod, external fuel tank, and a representation of a generic nuclear weapon. (Grumman History Center)

Although both were intended for high-speed, low-altitude ingress, the Buccaneer was more oriented toward high speed, and the Intruder, lowest altitude.

To facilitate the precision all-weather attack mission, the new A2F was to be equipped with an advanced system of radars and cockpit instrumentation called DIANE (Digital Integrated Attack/Navigation Equipment).[4] It had two radars, the Norden APQ-92 for ground mapping and the APQ-112 for acquiring and tracking the target. In addition to the usual TACAN, ADF, and radar altimeter capability, for accurate self-sufficient navigation it had an ASN-31 inertial navigation system, which used Doppler updates for increased accuracy. An early Litton digital computer with a drum memory integrated the radar and navigation input for the crew displays and weapons-control system. The pilot's primary displays were two cathode-ray tubes with a television-like presentation, one of the first applications in an aircraft. The weapons system provided an automatic weapons release on any radar-significant target for any reasonable dive angle, pull up g, wind, airspeed, and target range.

The mockup review was successfully accomplished in September 1958 and a contract for development followed. The originally proposed double-slotted slats had been changed to a simpler single-slot configuration and the wing trailing edge was straightened. The nose was enlarged to accommodate the search-and-track radars. The canopy was recontoured to reduce drag. The vertical fin was enlarged and moved farther aft for increased directional stability. The new catapult launch system with launch bar and holdback fitting integrated on the nose gear was incorporated.

The nose wheel braking was subsequently deleted to save weight. A cruise drag calculation error was also discovered, so the wingspan

Bob Smyth accomplished first flight of the basic A2F-1, less mission avionics and paint, in April 1960. The unusually large nose gear brace/retraction strut was required by the nose-tow catapult arrangement introduced with the A2F (A-6) and W2F (E-2). (Grumman History Center)

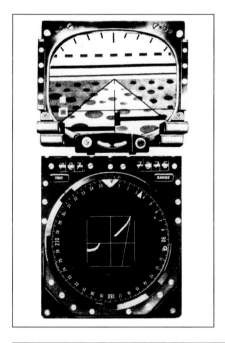

The upper instrument was the Visual Display Indicator (the dark spots that helped differentiate the sky from the ground were referred to as "cow flops"). The lower was the Horizontal Display.

tilting tail pipes did not result in a significant reduction in short field performance relative to their cost, weight, and complexity, so they (and the geared elevators) were deleted after the seventh production aircraft. The decision to have the fuselage-mounted speed brakes extend into the engine exhaust proved to cause buffet and handling quality problems. Perforating them and moving the horizontal tail farther aft mitigated these somewhat, but the eventual solution was bolting the fuselage speed brakes shut and adding wingtip-mounted ones. The rudder was enlarged for better spin recovery, with maximum deflection reduced with the flaps up and increased with flaps down.

The avionics suite was finally flown on the fourth A2F, which flew in December 1960. As might be expected, its development proved to require more time than hoped and delayed the introduction of the airplane into service. Terrain clearance mode wasn't attempted until the end of February 1961, and the Navy's first evaluation wasn't accomplished until November, almost a year after the system first flew, due to the lack of reliability and proper function. Deficiencies in brightness and resolution resulted in changes to both the pilot's and B/N's display hardware.

The A-6A pilot displays provided an all-weather and night capability for very low-level flight. Instead of the traditional mechanical artificial horizon, the pilot was provided with a large TV-like Vertical Display Indicator (VDI) with a contact-analog representation of the view forward of the airplane and the path to fly. It was a significant improvement over the use of multiple round-dial instruments for flight at night and in poor weather and all but essential for the task of safely controlling and maneuvering the airplane at very low altitude and high speed in hilly terrain. Susceptibility to vertigo in night and instrument flying conditions was significantly reduced. The display eventually included heading, radar altitude, vertical speed, angle of attack, and ILS data.

One terrain clearance feature was Search Radar Terrain Clearance (SRTC) using the APQ-92 search radar to generate a synthetic terrain display on the VDI. The presentation depicted vertical terrain within a 53-degree by 26-degree window centered on the projected flight path. For example, if the Intruder was heading for a valley between two hills, the pilot would see a representation of the hills, with a curving "V" notch between them. While SRTC allowed flight at night at low altitudes in hilly terrain, it did not "see" and therefore warn of hazardous features like cables strung across a valley.

Terrain data from the track radar could also be accessed by the pilot on his horizontal display, which was located directly below the VDI. The presentation, however, was a side view of the terrain ahead of the airplane, with the closest feature on the left of the display and the farthest on the right. A flight path line and clearance limit was also depicted. Since this was a different orientation than presented by the VDI, some mental gymnastics were necessary to correlate the two displays.

Terrain data was also presented on the B/N's primary reference, the Direct View Indicator (DVI), as a Plan Position Indicator (PPI) presentation covering an arc of about 50 degrees and a range of about 27 miles. It did not synthesize the terrain clearance data the way it was presented to the pilot but could be used to monitor his choice of flight path.

was increased by two feet for more aspect ratio, and the internal fuel capacity was also increased by 155 gallons in January 1959 to achieve the mission radius specification. In June, both the computer memory and the air-conditioning capacity had to be expanded based on a reevaluation of the avionics requirements, which were significantly greater than in past programs.

First flight was accomplished in April 1960 at Grumman's Calverton, Long Island facility. Neither this airplane nor the next two had mission avionics installed. Flight test was relatively trouble free but not without surprises that required configuration changes. The

Even though avionics system was still being developed, Board of Inspection and Survey (BIS) trials began in October 1962. Initial at-sea carrier-suitability trials were accomplished aboard *Enterprise* in December 1962, by which time the A2F had been redesignated A-6A. It and the E-2A Hawkeye, also aboard for the first time, were the first to be launched at-sea with the nose gear-mounted launch bar/hold back system. Trials were also accomplished from *Forrestal* in April 1963 and *Saratoga* in December 1963 and April 1964. Maximum catapulting and recovery gross weight was established as 58,000 and 37,000 pounds, respectively. Although it fell short in wind-over-deck requirements and longitudinal acceleration after launch, and there were other detail problems, the handling qualities in launch, wave-off, and recovery were satisfactory.

BIS was finally completed in January 1965, in part due to the lengthy development of the track radar system and the ballistic computer program.[5] The final BIS report was not complimentary:

> The service acceptance trials…have shown that the A-6A airplane has the potential to provide a significant improvement in fleet airborne/attack capability. However, the requirement to successfully attack and destroy moving targets under the cover of weather and darkness was not demonstrated as satisfactory during the trials. The difficulty in maintaining the complete attack navigation system and the lack of system reliability were major problem areas. The effects of rain and cloud conditions on both the search and track radars and the inadequacy of the terrain clearance provisions for protracted flight below 1,000 feet over mountainous terrain significantly reduced the specified all-weather capability. The excessive impact dispersion of system released weapons would prevent accomplishment of the pinpoint bombing requirement for isolated and moving targets during non-visual conditions.

Outside ring: In front of the A-6 are five mockups of Grumman multiple bomb racks. To the right and left are 46 Mk 81 250-lb bombs. The circle is closed in the back by 30 Mk 82 500-lb bombs.

Second ring: In front of the plane are six covered shapes marked "SECRET" that represent nuclear weapons. To the right are four 5-inch Zuni rockets followed by 13 LAU-10 rocket launchers. On the opposite side there are 19 2.75-inch rockets followed by 13 Aero 7D (LAU-3/A) rocket launchers. To the right and left of the horizontal stabilizers are 15 Mk 83 1,000-lb bombs. Twelve Mk 79 fire bombs are displayed behind the aircraft.

Inner ring: Behind the left wing (with the national insignia) are three Aero 8A practice bomb containers; on the opposite side, five Mk 84 2,000-lb bombs. In front of the left wing are five AGM-12A Bullpup air-to-ground missiles; in front of the right wing are four AIM-9B Sidewinder air-to-air missiles and a Douglas D-704 inflight refueling tank.

The aircraft itself carries five 300-gallon fuel tanks, with one not visible on the centerline station.

None of the weapon delivery accuracy specifications were met, with the level, "dive glide," and rocket modes using the computer-driven radar bombing system being particularly deficient. The A-6A was also judged to fall 5 to 18 percent short in guaranteed specific range. The empty weight was missed, but by only 200 pounds, less than 1 percent. Although the guarantees were not met, maximum mission radius with four 300-gallon external tanks and an Mk-28 store was impressive: 1,216 nm with 100 nm at low level on ingress.

Most concerning by far was the shortfall in reliability of the Integrated Attack Navigation System as measured in operating time between failure:

Subsystem	Requirement (Hrs)	Test Results (Hrs)
Search Radar	55	7.4
Track Radar	75	17.2
Ballistic Computer	20	14.3
Navigation	150	12.8
Optical Sight	610	333.0

Mission completion probability based on flights launched was 73 percent, and only 60 percent of scheduled flights were actually launched during service acceptance trials accomplished in early 1963.

Initial deliveries to the East Coast training squadron, VA-42, in February 1963 were not fully mission capable from an avionics standpoint either. The first airplane with a complete and qualified DIANE system was not received until June 1963 for training of the trainers. This delayed the induction of the first fleet squadron, VA-75, a former A-1 Skyraider squadron, until September.

The crew composition changed with the introduction of the Intruder. Instead of an officer as pilot and one or more enlisted men to operate the mission avionics has it had been since before World War II, the crew consisted of two officers, a pilot, and a bombardier/navigator, originally designated a Naval Air Observer and later a Naval Flight Officer.

VA-85 made the second and most troubled deployment of the A-6. The aircraft's avionics were not quite ready for prime time, and tactics for their effective utilization were still being developed. (Grumman History Center)

In 1966, camouflage was tried on attack aircraft in three air wings during combat operations in Vietnam. It was determined to provide no reduction in vulnerability compared to the standard gray/white paint schemes. (Robert L. Lawson collection)

When the system was working, a well-trained crew had the tools to launch, navigate to the target, attack, and return to the ship with no visual reference until final approach. Unfortunately, all elements of the system were rarely functioning properly at the same time during any flight. Pilots and bombardier/navigators had to be equally proficient at continuing the mission with more primitive methods and whatever information the system was capable of providing. A former enlisted B/N with A-3 ASB-1 experience was of significant value on those flights.

VA-75 deployed aboard *Independence* in May 1965 with 12 A-6s to participate in the escalation of the Vietnam War. They got off to a rocky start during the first month of strikes in North Vietnam; three A-6s were lost when a bomb on at least two, and probably all three, malfunctioned on release and blew up. The cause was determined to be the electric fusing of the bombs. A fourth aircraft was shot down in September.

VA-85 was the second Intruder squadron to deploy, departing on *Kitty Hawk* (CV-63) in late 1965. They fared even worse than VA-75, losing 10 of their original 12 airplanes. Initial results were disappointing in part due to the ongoing lack of reliability of the sophisticated DIANE system, exacerbated by the carrier operating environment and the hot/humid conditions of the Gulf of Tonkin. While the third A-6 squadron, VA-65, was preparing in early 1966 to deploy, the Secretary of Defense had begun to question the cost effectiveness of the A-6 versus the A-4 based on the results of the first two deployments. A-6 production was at risk of being terminated. Faced with that prospect, the Navy had the VA-65 airplanes sent back to Grumman for a tune-up of the DIANE system before their deployment in May 1966 aboard *Constellation*. VA-65 was also assigned many of the original personnel assigned to VA-42, so the experience level with the A-6 was much higher than with the first two squadrons.

Due to concern about unexplained losses in combat, a revised low altitude weapons-delivery technique, adding a steep turn after weapons release, was developed to minimize self-inflicted damage from general-purpose bombs. To further increase survivability and the element of surprise, crews planned tightly coordinated strikes at night and low altitude from different directions with the same time on target. For example, two A-6s simultaneously attacking the same target from opposite directions would release their conventional low-drag bombs from 800 feet and immediately make 60-degree right turns. The bombs would travel about a mile in the seven seconds between release and impact, which meant that the A-6s never got closer than ½ mile to each other, and both were clear of the resulting bomb blast.

Previously, major facilities such as power plants were attacked in daylight by large gaggles of airplanes. In such case, on 22 December 1965, a large electricity-generating plant was struck by 100 aircraft from *Enterprise*, *Kitty Hawk*, and *Ticonderoga*. The strike was effective, but two A-4s, one A-6, and one RA-5C were shot down, with four crewmen killed and two captured. In contrast, only three A-6s from VA-65 were involved in a subsequent and successful strike after dark in August 1966 with no losses.[6] VA-65 suffered no damage to, much less losses from, anti-aircraft defenses during night attacks; two were lost during day strikes. This deployment restored faith in the A-6

The A-6B cockpit was basically the same as the A-6A except for an additional pair of threat indicators, one for the pilot and one for the bombardier/navigator. (A viewing hood normally hid the B/N's radarscope.) (Grumman History Center)

program although use of the Intruder as a big A-4 in daylight missions in some air wings was not unheard of.

In early 1967, 10 A-6As were converted from the ground-attack mission to an *Iron Hand* capability by removing the track radar and ballistics computer and replacing them with radar detection, homing, and warning equipment. The A-6B was also given a new, bigger radar-seeking missile: the AGM-78 Standard ARM as well as the earlier AGM-145 Shrike. Like the Shrike, it was an adaptation of another missile, in this case the Navy's big RIM-66 Standard SAM. It had more range and speed, a bigger warhead, but used the same seeker as the Shrike. However, it also was 15 feet long and weighed 1,370 pounds. A subsequent seeker improvement increased the range of frequencies that a missile would home on, gimbaled so it had a wider field of view, and provided a memory function so it would guide to the vicinity of the radar antenna if the radar had been shut down. The first deployment was with VA-75 aboard *Kitty Hawk*, which departed the West Coast in November 1967 en route to Vietnam. Two or three were to be assigned to the 12-plane A-6 squadron on each carrier to provide SAM suppression during strikes. The A-6B could carry two less expensive Shrikes, two Standards, and an external tank.

The first 10 A-6Bs were followed by three conversions that retained the A-6's tracking radar and incorporated the Johns Hopkins-developed Passive Angle Tracking (PAT-ARM) system. This combined the ranging function of the tracking radar with passive ECM antennas that aimed it at the enemy site they detected to provide accurate range and bearing data to the ballistics computer. An additional six modified in 1970 with the IBM APS-118 Target

The A-6C was configured with a huge, self-contained sensor package on the belly that provided infrared and low-light television-generated images to the pilot and B/N. While not fully satisfactory, it provided combat experience with the technology. The hardware was subsequently reduced in size from a telephone booth to a television set with equally dramatic, if not greater, improvement in picture quality. (Grumman History Center)

Identification Acquisition System (TIAS). It more completely integrated the functions of the onboard systems and the AGM-78 itself to maximize the standoff range of the ARM.

Even the A-6A's big radars had limitations with respect to detection of relatively small objects in ground clutter, like the trucks that were resupplying the North Vietnamese troops. In order to see better in the dark, 12 A-6As were modified to carry a large, 3,000-pound pod on the belly that contained both an infrared (IR) sensor and a low-light level camera, along with all the associated avionics and power supplies. These were designated A-6C. The infrared sensor depicted the view in terms of temperature differences, requiring no light whatsoever, while the low-light level camera enhanced the picture but required a minimal amount of light. The A-6C avionics included Black Crow, which detected and indicated the direction to truck engine ignition systems. All together, the system was known as TRIM: Trails, Roads, Interdiction Multi-sensor. VA-165 was issued some of the A-6Cs along with As and Bs for a 16-plane deployment that began in April 1970 aboard *America* (CV-66). TRIM received mixed reviews from an effectiveness standpoint, with its weight, drag, and reliability being notable drawbacks. Its primary benefit was to provide a combat evaluation of the relative merits of IR and low-light level sensors. No additional aircraft were modified, but various squadrons operated the A-6Cs until their turn came to be converted into A-6Es.

Survivability in the Vietnam War had begun to depend more and more on jamming the enemy's radar and communications. The Marines used converted and obsolete Douglas EF-10 (F3D) Skyknights while waiting for an A-6 modified for the mission. Grumman began work on the A2F-1Q in August 1961. Redesignated EA-6A, the first one flew in February 1963. Only 28 were built,

including three prototypes and 10 airplanes modified from A-6As. All were delivered to the Marine Corps with the first ones going to Vietnam in 1966. (Some were operated off carriers by Marine Corps detachments in the early 1970s to provide a bridge between the retirement of EA-1Fs and the introduction of the EA-6Bs.) The EA-6A configuration included a pod on the tail housing antennae for radar and communication signal reception and a stores station on each outboard wing panel. Different jamming pods, chaff pods, and Shrike missiles could be carried depending on the mission and threat systems.

Significantly more capability was built into the EA-6B Prowler, which added two more crew positions in a stretched forward fuselage. The increased gross weight required higher-thrust J52 engines. The first one flew in May 1968. It carried the ALQ-99 tactical jamming system. The Prowlers retained the five stores stations to carry a mix of jammer pods, fuel tanks, and/or anti-radiation missiles depending on mission requirements. Each jammer pod is integrally powered and houses two jamming transmitters that cover one of seven frequency bands. The first production airplane was delivered in January 1971 with the first deployment of a squadron, VAQ-132, in June. The ALQ-99 was upgraded every few years to have more automation and performance in detecting and jamming enemy transmission. Surviving aircraft were modified to the newest configuration. Even though some of these upgrades were major changes, none ever resulted in an EA-6C designation. Only 170 EA-6Bs were built, with the last one delivered in 1991. Continuous avionics systems updates and replacement of wing center sections (a few aircraft more than once) and outboard wing panels are supposed to keep the Prowlers effective and deployable through at least 2015, making it the longest lived of the Navy's attack airplanes.

The A-6B was dedicated to the Iron Hand anti-radar mission. This is an early version, carrying two Standard ARM missiles on a test flight from San Diego International Airport. (Grumman History Center)

Making do with what's available, the Marines had some of its obsolete Douglas F-10 (F3D) night fighters converted for electronic warfare missions to cover strikes in Vietnam. The EF-10 was a placeholder until Grumman and the Navy could complete the development and qualification of the EA-6A, which was tailored to the electronic warfare mission. This VMCJ-1 EF-10B is marked as a 16-mission veteran. (Terry Panopalis collection)

Although the KA-3 Whales were excellent tankers, they were wearing out, and it made sense to minimize the number of different airplane types in an air group from a maintenance standpoint. The result was the conversion of A-6A bombers into dedicated tankers, the KA-6D, with the radars and mission avionics removed and an integral hose reel and drogue system built into the aft fuselage. With five 300-gallon external tanks, the KA-6D could give away 26,000 pounds of fuel. The first KA-6D conversion flew in April 1970. The first deployment with VA-115 aboard *Midway* began a year later. An A-6 squadron would typically have four to five tankers assigned in addition to nine or ten bombers.

In the late 1960s, the Navy began to address the upgrade of the A-6 avionics suite. The result was the A-6E. The two different radars for search and track were replaced with a single multi-mode radar, the APQ-156, retaining all the same functions and extending detection range. DIANE was replaced with an all-digital system that incorporated a better computer and a centralized armament control unit. The first A-6E flew in February 1970, with the first production aircraft delivered in September 1971. The incorporation of built-in test equipment and the addition of an onboard avionics cooling system improved reliability over the A-6A, but were initially short of the

Carrier qualification of the Grumman EA-6A was accomplished aboard Kitty Hawk in 1966. Note the extra pylon on the outboard wing panel and the antenna pod at the top of the vertical fin. The nose equipment bay was extended by eight inches to provide more volume and balance the tail-mounted pod. The pod on the outboard wing pylons is the ALQ-53 for ECM reconnaissance, installed only on the first dozen EA-6As. The ALQ-31 jammer pods on the middle wing pylons date back to the AD-5Q and were soon replaced by the ALQ-76. (Grumman History Center)

The two-place EA-6A, which was used exclusively by the Marine Corps except for a Navy aggressor squadron, was followed by the bigger and more capable four-place Grumman EA-6B. Because of the additional weight of the fuselage stretch, the EA-6B did not have the extra pylons under the outboard wing panels like the EA-6A. (Grumman History Center)

The A-6 also provided an excellent dedicated tanker, here refueling an F-4J Phantom, a regular customer. Although a buddy pod could be used to temporarily convert any A-6 into a tanker, the KA-6s like the one shown here had been permanently modified for the tanking role with an internal hose reel in place of the mission avionics. It often carried a buddy tank on the centerline station as well as a backup for the internal package. (Robert L. Lawson collection)

The EA-6B jamming capability could be tailored to the mission by the mix of ALQ-99 jamming pods carried. The propeller on the nose of the pod turns a generator, which provides electrical power to the pod's transmitters. It was replaced by an antenna on the inboard wing leading edge. (Jay Miller collection)

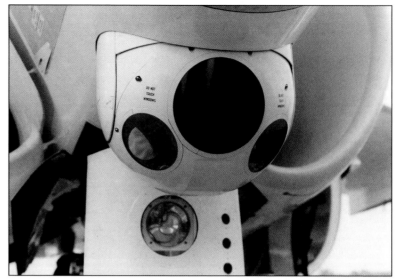

The A-6's TRAM turret could rotate 360 degrees and pivot up and down. The infrared sensor was located behind the big center window and incorporated a zoom capability. The smaller window on the left was for the laser designator and rangefinder. The one on the right was for the laser spot detector, which allowed the B/N to see a target being designated by a forward observer or another aircraft. (Grumman History Center)

desired levels. Bombing accuracy also improved. In the mid-1970s, the INS was replaced with CAINS, Carrier Airborne Inertial Navigation System, which provided a quicker alignment prior to launch.

The next upgrade to the A-6E took advantage of the development of laser-guided weapons, advances in night-vision technology, and the space made available in the Intruder's big nose by the deletion of the search radar.

The laser-guided bomb had been pioneered by the Air Force in the late 1960s. It was a simple and relatively inexpensive conversion of existing bombs by the addition of a laser seeker head and a tail cone with control fins. The target was designated by a laser spot and seeker head input was used by a computer to move the fins so the bomb tracked to the spot. The initial aim of the bomb only had to be approximate; if the electronics didn't fail, the bomb would hit within 10 feet of the target. Designated by the Air Force as Paveway (Precision Avionics Vectoring Equipment), it was first used in Vietnam in May 1968. A few A-6As were fitted with the Air Force developed *Pave Knife* pod, which provided laser designation for *Paveway I* laser-guided bombs. VA-145 deployed with the pod in November 1972 aboard *Ranger*. Like the A-6Cs, the *Pave Knife* airplanes were passed from squadron to squadron for a while for laser-designation familiarization.

By 1972, FLIR (Forward Looking Infrared) had been selected as the preferred nighttime seeker for use in attack aircraft over illuminated television (ITV), which enhanced the sensitivity of a low-light-level television camera with an infrared illuminating flash. Although ITV provided a better picture in some circumstances, FLIR was ade-

quate and not detectable since it was passive, simply registering the difference in temperature of objects. ITV illumination could be seen on infrared detectors and provided the basis for counter fire. The result was a gyro-stabilized chin-turret containing a FLIR, laser range-finder, and a laser target designator. It was referred to as TRAM, Target Recognition Attack Multi-sensor. The first TRAM equipped aircraft flew in March 1974. The first full-systems aircraft wasn't delivered to VA-142 until December 1978. The last A-6E was completed in January 1992 and delivered in April.

FLIR and the laser-guided bomb had limitations. Clouds, haze, humidity, smoke, and dust diminished the bombardier's ability to detect, identify, and designate targets. The display was not as crisp as the B/N would have liked. However, with a clear line of sight to the target, a laser-guided bomb attack was far more accurate than one on radar alone.[7]

The final A-6 major upgrade was the Systems/Weapons Improvement Program (SWIP). This was to include a new composite wing as well as incorporate an upgraded ECM suite, allow the use of night-vision goggles, and add the capability to carry the latest Maverick, SLAM, Sidewinder, and High-speed Anti Radiation Missiles (HARMs).

A total of 488 A-6As (including the prototypes) were built before production switched over to the A-6E version in December 1970. Of these, 19 A-6As were converted to A-6Bs, 12 to A-6Cs, 90 became KA-6D tankers (12 after they had been A-6Es), 13 were converted to EA-6As (there were also 15 new-build EA-6As), and three became

The A-6E has just dropped a CBU-59 cluster bomb in combat on 18 April 1987 and it has split open to release its bomblets. The cluster bomb replaced napalm and fragmentation bombs for missions requiring suppression of enemy forces over a large area. The cluster bomb containers could be loaded with different bomblets effective against specific targets like armor or personnel. (U.S. Navy DN-SN-89-03124)

the prototypes for the EA-6Bs. About half of the A-6As (240 in all) were later upgraded to be A-6Es. New A-6Es included 95 built before TRAM became available, 71 TRAM, and 41 SWIP. Of all the A-6Es, 188 eventually were SWIPed, of which 136 received composite wings.

Standoff Weapons

The AGM-84 Harpoon was developed in the early 1970s as an all-weather, low-level, over-the-horizon, anti-ship missile. It was powered by a small jet engine and guided by an autopilot to a previously designated location where the onboard radar acquires a target for the attack. The surface-to-air version was 12 feet 7.5 inches long and weighed 1,145 pounds with a range in excess of 60 nm. The completion of its operational evaluation was a VX-5 launch of a Harpoon with a live 500-pound warhead from an A-6E at a destroyer target in 1980. It was deployed in 1981. The first in-anger use of the Harpoon was in March 1986 off Libya, south of Colonel Ghadhafi's "Line of Death." A-6Es sank a missile-equipped patrol boat that appeared to be targeting *Ticonderoga* and damaged another one. The next day, another patrol boat was sunk and one damaged.

The main shortcoming of the Harpoon was its indifference as to whether the target it homed in on was the one intended. In December 1988, a Harpoon launched from an F/A-18A in a training exercise at sea near Hawaii struck an Indian merchant ship that was in the exercise area instead of the intended target. Although the Harpoon wasn't armed, one of the ship's crew was killed.

The AGM-84E Standoff Land Attack Missile (SLAM) was developed from the Harpoon airframe in the late 1980s by removing the radar and adding a GPS receiver, Maverick infrared imaging seeker, and Walleye datalink. The missile can either be programmed to hit a specific set of coordinates or once in the terminal area, the controlling airplane can view the seeker image and lock the missile onto a specific target. It is slightly longer and heavier than the Harpoon at 14 feet 9 inches and 1,385 pounds, respectively. Range is about 60 nm. In 1990, VA-34 aboard *Eisenhower* was the first squadron to deploy with SLAM. It was first used operationally in Desert Storm in January 1991 to successfully attack a power-generating facility without damaging an adjacent dam or endangering the strike pilots.

No Longer Built From Iron

The A-6A began to have structural problems in 1969 with the aircraft being operated at heavier weights and higher load factors more often than originally assumed. One crew was lost in the U.S. in August when a wing failed during a practice bombing run. In February 1970, a wing came off an airplane flying a mission over Laos, with the crew ejecting successfully. However, another wing failure killed the crew in June. An inspection and repair program was initiated. Replacement of wings on older aircraft became part of the overhaul program that converted A-6As to A-6Es.

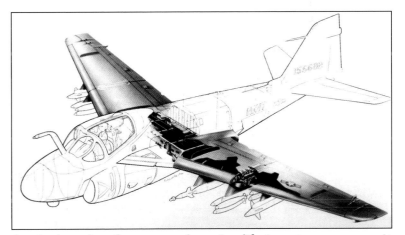

Rewinging aircraft to extend service life is not uncommon. In the case of the A-6, Boeing designed a new wing under contract to the Navy that was primarily built using graphite composite material instead of metal, providing a higher strength for the weight and virtually no susceptibility to corrosion. Note the extra stores pylon to be incorporated in the A-6F configuration. (U.S. Navy via author's collection)

The Air Force equivalent of the A-6 Intruder was the F-111 Aardvark. It was developed and first fielded in the 1960s. It incorporated a variable-sweep wing and afterburning turbofan engines to provide both transatlantic range for self deployability and low-altitude supersonic speed for ingress. This F-111F is armed with four Mk 84 laser-guided bombs. A Pave Tack infrared sensor and laser designator is mounted in what was originally a bomb bay for a nuclear weapon. Maximum gross weight was almost 100,000 pounds. (Terry Panopalis)

In 1985, as part of an A-6E Systems/Weapons Improvement Program (SWIP), Boeing received a contract for a composite A-6 wing. To minimize flight qualification effort, it was to have the same external shape of the original wing. With the benefit of carbon fiber structural characteristics, it was to have a fatigue life of 8,800 hours, twice that of Grumman's warranty on its metal wing, and be virtually impervious to corrosion. It was none too soon, with another fatality in January 1987 following wing failure during dive-bombing practice. At one point, half the fleet of 344 A-6Es was grounded or operating under maneuver restrictions.

Design and fabrication of a composite wing proved more challenging than Boeing or the Navy had anticipated. Wind tunnel testing of a scale-model wing with the stiffness characteristics of the new design resulted in catastrophic flutter and required redesign and retest in June 1987. Manufacturing problems were encountered as well. Nevertheless, in May 1988 the first wing was installed on an A-6E at Grumman, and as of mid-1989 Boeing had a contract for 179 wing sets, with options for 148 more. The second wing was installed on the flight-test aircraft at Boeing Wichita. It first flew in April 1989 and was subsequently ferried to NATC for flight evaluation.

The first A-6E with a composite wing was finally delivered to the fleet in October 1990, two years behind the original schedule. Due to delays in the Boeing program, the first 15 of the 39 new-build A-6E (SWIP) airplanes did not have the composite wing. In the end, only about half the fleet was retrofitted due to a budget-driven decision by the Navy in 1993 to retire the A-6 much earlier than previously planned.[8] All of the A-6 upgrades, including the Boeing rewing program and a Block 1A improvement to the SWIP configuration which had flown for the first time in September 1992, were terminated. The number of deployable A-6 squadrons was steadily reduced with the last A-6E bomber deployment ending in December 1996 when VA-75 flew off *Enterprise*. Fittingly, they had been the first squadron to deploy with the A-6 as well, more than 30 years earlier.

It proved hard to replace.

Skipper II was another innovative China Lake creation to meet a 1980 request for more standoff distance. It was created from an Mk 83 LGB by adding a Shrike rocket motor and was fielded in 1985. It had a maximum range of 30 nm, but the effective range was limited to about half that by the requirement for laser designation of the target. It was phased out in the early 1990s in favor of the laser Maverick. (Norm Filer)

Dropping bombs at low altitude risked, as Shakespeare noted, having "the engineer hoist with his owne petar," the petard being a 16th-century military engine, now known as a bomb, placed against a gate or wall to force a breach. The Snakeye was an Mk 80-series bomb modified with a high drag device in place of the standard afterbody and fins. It automatically popped open after release. This resulted in the ability to make a low-level, high-accuracy, lay-down drop. The Snakeye was qualified in 1963 and fielded in 1964. It could also be dropped as a conventional bomb with the fins locked shut. (Robert L. Lawson collection)

The A-6 was to be armed with the ASM-N-11/AGM-53 Condor. This was a TV-guided, rocket-propelled missile with a range of approximately 60 nm for high-precision standoff attacks. Rockwell received a contract in July 1966. First flight was in March 1970, with part of the delay caused by a change from a liquid to solid rocket propulsion. Development and operational evaluation was completed in 1975, but the missile proved too expensive and unreliable so no production resulted. (Terry Panopalis collection)

The Harpoon was initially developed as an antiship missile, shown here on an A-6 during a test flight in 1983. It has seen only limited, albeit effective, operational use in that role. The design was subsequently modified to be the Standoff Land Attack Missile, SLAM. (U.S. Navy DN-SC-83-06997)

CHAPTER NINE

LIGHT ATTACK REDUX

In 1966, the A-6A had replaced the AD-5N Skyraider for all-weather attack, but the single-seat AD-6 was still serving alongside the Intruder, as shown here on Kitty Hawk, because of its load-carrying capability and endurance compared to light jets. Its replacement, the Vought A-7A Corsair II, was just beginning to be produced in time for the aging and obsolete ADs to be retired, at least from the carrier strike force, in 1968. (U.S. Navy via author's collection)

In 1960, the Navy's project for replacement of the single-seat AD Skyraider and the A4D Skyhawk was known as VAX, with V standing for heavier-than-air; A, attack; and X, experimental. In addition to more payload and range, it was to be supersonic even at sea level, in keeping with the presumption that high speed reduced the likelihood of detection in time for an effective defensive action. These preliminary requirements were provided to the Navy's aircraft sup-

pliers, each responding with design studies, mostly with two engines and variable-sweep wings. At that point, the Navy would normally have held a competition for a new aircraft or even awarded a contract without a competition.

In early 1961, however, Robert McNamara became the Secretary of Defense in the Kennedy administration. Among other things, McNamara instituted systems analysis and increased inter-service commonality in procurement, significantly changing the relationship between the Office of the Secretary of Defense and the military services. He also expanded OSD oversight into the services planning and decision making process, requiring competitive procurements that were reviewed up through his office, effectively ending the more informal approach, particularly in the Navy, that sometimes prevailed. The Vought Corsair II was one of the more successful programs accomplished during his tenure.[1]

In mid-1962, McNamara suggested that the Navy consider joining with the Army and the Air Force in a tri-service program for an advanced light attack aircraft. He followed this up in November 1962 with an offer to fund VAX after the Navy quantitatively validated the operational requirements. This resulted in the Navy forming the Sea Based Air Strike Forces Study Group, whose efforts encompassed all carrier-based aircraft attack and fighter types as well as V/STOL concepts. In support of this study, BuWeps evaluated and priced out 144 different aircraft programs including avionics variations. Measures of effectiveness were used to compare and contrast the options. Self-defense, i.e., air-to-air capability, was a consideration along with the typical close-air support and interdiction (denying the enemy resupply of their troops) missions.

The surprising conclusion was that the supersonic capability was not worth, in terms of survivability and weight of bombs on target, either its development cost or unit price. The Navy estimated that it could develop and buy three times the number of subsonic attack aircraft for the total program cost of a supersonic one. From a survivability standpoint, the drag of externally loaded bombs, the only practical way to carry a full payload of conventional (non-nuclear) weapons, precluded supersonic flight on the way to the target. Even if he could fly faster, the pilot would probably not be able to fly as low as the pilot of the slower airplane, which made vulnerability of the two roughly equal.

Although near-term operational availability was desired, the Navy wanted the new airplane to be powered by its new Pratt & Whitney TF30 turbofan engine. More efficient than existing axial flow engines, it would allow a 600-nm radius of action with a greater load of conventional bombs.[2] For their part, OSD stipulated that the development contract be fixed-price with cost, schedule, weight, performance, and reliability/maintainability guarantees backed by penalties for failure to perform.

The Specific Operational Requirement for a subsonic light attack airplane, VA(L), renamed to distinguish it from the supersonic VAX program, was released by the Chief of Naval Operations in May 1963. The requirements were provided to industry a week later and a request for quotation issued by BuWeps at the end of June to Douglas, North American, LTV, and Grumman. Since a fixed-price contract was only appropriate if the development risks were low, the BuWeps Request for Quotation stated that "VA(L) must be a modification of an aircraft currently in the Navy inventory." The use of the TF30 turbofan engine was also specified. Technical proposals were required in less than six weeks with cost proposals to be submitted three weeks later.[3]

The primary attack mission was the delivery of six Mk 81

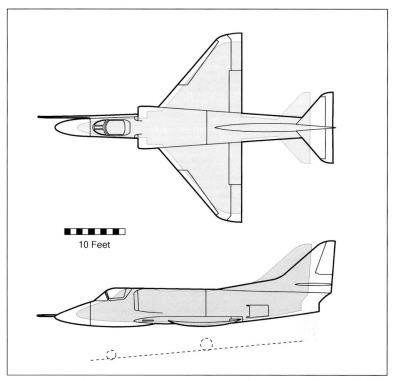

Figure 9-1. The A4D-6 was a 1962 Douglas proposal to scale up its Skyhawk for more payload and range, substituting the new, more fuel-efficient turbofan TF30 engine for the A4D-5's axial-flow J52. The Navy held a competition and selected the Vought Attack Crusader proposal instead. (Author)

Snakeye bombs at a radius of 600 nm. The overload mission was 12 Mk 81s at the same distance. The land-based close air support mission was to lug 7,500 pounds of bombs to a radius of 200 nm and loiter for one hour. (This was actually 32 percent more bombs and 100 percent more time on station than the A-1H Skyraider could provide at a 200-nm radius.) Maximum bomb load capacity was to be 12,200 pounds across six stations and as many as 26 bombs. The airplane was to be armed with two 20mm cannons with 250 rounds of ammunition per gun. Only limited all-weather capability was required, and weapons delivery was to be visual only. Significant emphasis was placed on reliability and maintainability.

Grumman management carefully weighed the competitive prospects of both a TF30-powered F11F Tiger and a single-seat derivative of the A-6 because they had to choose between being compliant with the engine requirement versus paying more than lip service to the modification of an existing airplane. They decided to promote the commonality benefits of a single-seat A-6. The complex avionics system was replaced by one incorporating a simpler multi-mode radar. Folding was added to the horizontal tail to increase the number that could be stowed in a given area. A 20mm cannon was added on each side of the nose. Since the required payload-range was not a problem for the big A-6, it retained its J52 engines to reduce development cost and increase

Grumman's single-seat VA(L) proposal was by far the least extensive modification to an existing airplane of those submitted. An inverted-T adapter provided for the minimum six pylons required by the RFP specification. A 20mm cannon was also added to either side of the nose. (Grumman History Center)

A wing pylon with air-to-ground stores capability was added to the Crusader beginning with the F8U-2. In February 1964, CDR Joseph Simons, Commanding Officer of VF-162, dropped the first bombs, albeit Mk 76 practice ones, from a deployed F-8 Crusader while flying from Shangri-La. (U.S. Navy)

The North American FJ-4B Fury had been out of production since May 1958 but it had been well regarded. Although the last fleet squadron had turned theirs in for A4Ds the year before the VA(L) competition, it was still being flown in the reserves. The requisite radar was added under the enlarged engine inlet in order to not further compromise the all-important over-the-nose visibility. The main gear retracted forward into fairings that extended forward of the wing leading edge to increase the volume available in the wing for fuel. (Tony Buttler collection)

commonality with the two-seat A-6s in the air wing. There was the problem of the higher unit cost of a bigger airplane, but Grumman could claim a lower development cost, significantly better payload-range than required, earlier operational availability, and operating cost savings and benefits because of its high commonality of engine, structures, and systems with the all-weather A-6s in the carrier's air wing.

The other three competitors proposed derivatives of their existing airplanes reengined with the TF30 turbofan engine: Douglas, the enlarged A-4; North American, a modification of the FJ-4 Fury; and Vought, a major redesign of its F8U Crusader. Douglas was considered to be the odds-on favorite to win the competition based on its incumbency and familiarity with the mission. As described by a Standard Aircraft Characteristics chart dated 1 August 1962, the so-called A4D-6 was similar to the A4D-5, then in production, and retained the same five stores stations. However, as shown in Figure 9-1, the proposed airplane was the A4D on steroids, bigger in every dimension to accommodate and take advantage of the bigger TF30, which had almost 3,000 pounds more thrust than the A4D-5's J52. Internal fuel was increased by 290 gallons (36 percent). As a result, while retaining the overall look of the Skyhawk, it was significantly different in detail, particularly the structure. With more fuel and the efficient TF30, it had a radius of action with an Mk 43 and two 300-gallon tanks of 960 nm, 52 percent more the A4D-5's.

Vought management recognized that their prospects of winning would require a superior design, the lowest cost, and willingness to agree to the onerous, OSD-dictated contract terms. Attack experience was a concern. Douglas had the most. Vought's was limited to derivatives of their fighters. For example, an air-to-ground capability had

Officially, there wasn't a bomb shortage in Vietnam but sometimes the ordnance team had to go back pretty far in the bomb dump to find something to hang on the aircraft. This VMF(AW)-235 F-8E is taking off with a World War II 2,000-lb bomb under its right wing in addition to the two two-tube Zuni rocket launchers mounted on the side of the fuselage. (U.S. Marine Corps)

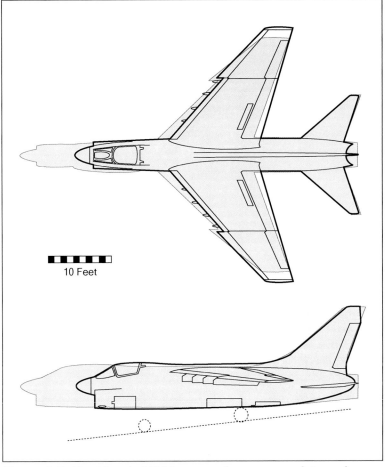

10 Feet

Figure 9-2. The Vought VA(L) proposal was a complete makeover of its F8U Crusader to otherwise fully respond to the Request For Quotation. Since there was no afterburner and only two guns, the forward fuselage had to be shortened to maintain the required center of gravity. The single speed brake of the Crusader was replaced by three: one on either side of the lower forward fuselage and one aft of the main landing gear wells. The upper aft end of the vertical fin was cut back in a late drawing board review of the layout to further shorten the airplane and maximize the number that could be parked on the carrier. (Author)

just been added to the F8U-2NE model of the Crusader, including provisions for the Bullpup missile.[4] It was in production in 1963 as the F-8E. The two fuselage-mounted pylons were adapted to carry Zuni rocket pods. Two wing pylons were added that could carry up to 2,000 pounds each of ordnance or fuel tanks.

Although ostensibly a modification of the F-8 and resembling it as shown in Figure 9-2, the proposed "Attack Crusader" had little in common with it. The structure was almost all new. For example, the wing was bigger with different airfoils and no longer had variable incidence. The ailerons were moved to the outboard panels and three weapons pylons were provided on each inboard panel. The fuselage was almost 10 feet shorter and the cockpit was 3.5 inches wider to provide more console space. The pilot was also positioned higher for better visibility over the side and forward. The fuselage break for engine removal was modified to simplify the engine change process and reduce weight. The flight-control system was revised to eliminate the F-8's handling quality problems on approach. Range was assured by adding 11 percent more internal fuel to the already enormous, for a fighter, F-8 fuel capacity. To maximize the number that could be accommodated on the carrier, Vought proposed a folding horizontal tail option.

Avionics included a navigation computer, an APQ-99 multi-mode radar, electronic countermeasures, and automatic weapons delivery system. A separate Doppler radar measured airplane ground speed and drift angle to provide navigation data to the navigation computer. The computer combined the Doppler data with a known starting position, true airspeed, and heading changes to maintain a continuous present position on a roller map. One small TV-like display was used

to provide ground mapping, terrain following and avoidance, air-to-ground ranging, ECM information, and TV from a Walleye missile. The automatic weapons-delivery system used input from the radar, air data computer, AHRS (attitude and heading reference system), angle-of-attack sensor, and an accelerometer to provide signals to the pilot for weapons delivery.

Vought's brochure creativity and salesmanship went into overdrive with respect to the issue of its proposed airplane being a modification of the F-8. Big pie charts were used for systems components and showed that the percentage of new system components was minimal. Close examination showed that the majority of the non-new

Only six weapons pylons were required by the VA(L) RFQ, but Vought management decided that 10 would provide a competitive advantage. As it turned out, the Navy immediately questioned the viability of stores clearance, particularly if jettison was required. (Vought Aircraft Heritage Foundation)

Stores release/jettison was typically evaluated first in wind-tunnel tests to reduce the risk of an unsatisfactory result during the required flight-test demonstration. This multiple-exposure photo of a minimum-release-interval drop of bombs from the middle pylon shows their proximity to the bombs on the outboard pylon. These tests were used to predict how released stores would behave in the presence of the airflow around the aircraft and adjacent stores. (Vought Aircraft Heritage Foundation)

components were "similar or off-the-shelf," not common with the F-8E. In smaller print, the proposal brochure revealed that the commonality of structure was only 13 percent. This turned out not to be a disadvantage, since all the other competitors besides Grumman were also proposing major changes to existing airplanes and Grumman's proposal was too expensive.

In BuWep's assessment, the North American and Vought proposals were considered to have about the same payload range, with the Vought proposal being lower in cost. The Douglas proposal was about the same cost as Vought's, but had less payload and range.[5] The Grumman proposal doesn't appear to have made the short list. Although "Vought was a hands-down winner," according to George Spangenberg in his oral history, the decision had to be justified to OSD with a very extensive and detailed analysis:

> We must have done thirty different missions from six Mark-81s, twelve Mark-82s, from little payloads, every combination you can think of, Rockeyes, Walleyes, everything. All grossly not needed by anyone experienced in the art. You took a look at the payload-range curve and you knew what all the answers were going to be but we had to reduce it down and show the bombs per buck or something to get the whole job done. We did more radius work in that competition than we had in all the competitions we ever had before put together. Drove us crazy.

VADM John S. Thatch, the creator of the Thatch Weave fighter tactic in World War II, was the Deputy Chief of Naval Operations for Air Warfare from 1963 to 1965:

> We specified that the bids by each prospective contractor (be fixed price) rather than being cost plus fixed fee…. And rather than having bonuses for meeting requirements or exceeding them, … we said there was going to be a penalty for not meeting the requirements in each specific area such as speed, various performances, rate of climb, time to produce a certain number of airplanes, the man hours required to maintain the airplane, the number of hours in commission versus number in repair. In many areas, we asked the contractor to set the amount of the penalty. What did this do? It separated the men from the boys, the people who knew how to build airplanes and the people who may not and do not have confidence.
>
> This was the first time we had tried such a thing and it was most interesting to me to find the difference between the amounts of the penalties the various companies proposed. In other words, one of them might propose a million dollars for

The number of A-7 wing pylons had been reduced to three on each side prior to the mockup review in June 1964. The three speed brakes had also been reduced to one under the fuselage like the Crusader's. The radome still has a sharp nose. This would be blunted to minimize the impact of a fuselage stretch to position the center of gravity properly. (Vought Aircraft Heritage Foundation)

failure to meet the production schedule, or take a penalty of $750,000 if the rate of climb is a certain amount below what they guarantee. We told the contractors that we want guarantees and we want them backed up by a figure that you're willing to accept as a penalty. In other words, how confident are you.

Well, there was a wide variance in how confident they were. Some went way up in some areas and weren't willing to take much of a penalty with other specifications. So it really put the spotlight on where their weaknesses were.

Not only were the proposal and the design of the LTV airplane better, but they put up fantastic penalties for failure to do it. They were confident that they could do it and they considered it a very low risk program and were willing to bet their shirts.[5]

Although the Navy had decided on Vought as the contractor for the A-7, completed its justification, and had OSD concurrence before Thanksgiving 1963, the requisite review, approvals, and budget action by Congress resulted in the decision not being announced until February 1964. Vought finally received a contract for the A-7A on 19 March.

A-7A Development

While not really a modification of an existing airplane, the A-7A was still a low-risk program except possibly for its engine, justifying the use of a fixed-price contract. The avionics were basically off-the-shelf. The airframe and planform were conventional, at least similar to the F-8's, and simpler than the F-8's. For once, Vought engineers had foregone their penchant for the innovative and complex and designed a relatively simple and maintainable airplane. However, the engine was not only a new design, not flying for the first time until December 1964, but was a new type. This was to cause the only hiccup in the A-7's otherwise quick and smooth transition to operational use from the mockup review in June 1964 to dropping bombs in combat in Vietnam on 3 December 1967, three and a half years later.

First flight on 27 September 1965 was accomplished ahead of a challenging schedule, cost overruns were not significant, and Board of Inspection and Survey trials and the Fleet Indoctrination Program began only a year later. While these were being completed, the first production A-7As were delivered to the Navy training squadron on

The third A-7A prototype was used to demonstrate that it could carry a lot of bombs, in this case 26 Mk 81s for a total bomb load of about 6,500 pounds. This was by no means the limit. During A-7E carrier qualification trials, at least one launch was made with 19,000 pounds of external stores, including pylons and MERs. (Vought Aircraft Heritage Foundation)

The only significant hiccup in A-7 development was, as usual, unexpected. The low-mounted engine inlet sucked in steam escaping from the catapult track. The ingestion of extremely hot air had a deleterious impact on the turbofan engine's compression cycle, resulting in momentary compressor stalls and loss of thrust during launch. This was mitigated by a combination of changes to the engine and the catapult track sealing. (Vought Aircraft Heritage Foundation)

146

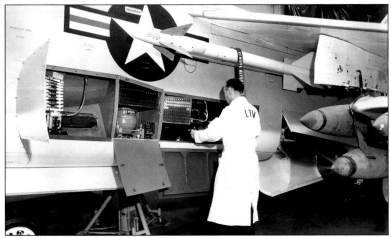

As demonstrated on the mockup, A-7 maintainability was enhanced by large, waist-high doors on the side of the fuselage. From front to back, the compartments contain the 20mm cannon and ammunition feed chute, the oxygen bottle, and the avionics. The avionics were grouped together in subsystems and readily accessible behind quick-opening doors. (Vought Aircraft Heritage Foundation)

the West Coast, VA-122, in November 1966. (VA-122 was still training pilots in the A-1 Skyraider at the time.) By the time the final Service Acceptance Trials Report had been issued by the Board of Inspection and Survey on 2 August 1967, over 100 A-7s had been provisionally accepted.

Weight empty was the only missed guarantee, primarily due to a management decision to accept a heavier wing structure than needed initially in expectation of future growth. According to Vought, the cost of the penalty was worth the potential business that would result. The maintainability requirement, 17 maintenance man-hours per flight hour, was probably one of the most challenging given that it was historically 30 to 40, but Vought demonstrated it successfully.[7] The requirement to deliver 6 Mk 81 bombs at a radius of 600 nm with only internal fuel was bettered—the A-7A could deliver 16 Mk 81s to a radius of 600 nm and the required 6 to 700.

One particularly welcome attribute was that the A-7A could land back aboard with as much as 6,000 pounds of fuel. Since only 300 pounds were required for a circuit around the carrier in a six-plane pattern, this was a comfort level previously provided only by propeller-driven airplanes.

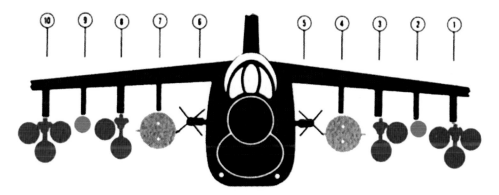

AIR-TO-SURFACE GUNS AND ROCKETS ARRANGEMENT

STA 1, 3	10-LAU-3/A 2.75 FFAR PODS OR 10-LAU-10/A 5.00 FFAR PODS TOTAL BOTH STATIONS
STA 8, 10	SAME AS STATION 1 & 3
STA 4, 7	2-MK 4 GUN PODS EACH STATION
STA 2, 9	2-LAU-32A/A 2.75 FFAR PODS EACH STATION

Although more weapons pylons seemed like a good idea, there were very few combinations of weapons that could be loaded on pylons placed so close together. This is the only one of six alternatives depicted in the Vought proposal brochure that has a load on every pylon. (Vought Aircraft Heritage Foundation)

The BIS report compared the A-7A to the A-4E that it was to replace:

	A-4E	A-7A
Wingspan Folded (ft)	27.5 (no fold)	23.8
Length (ft)	41.3	46.1
Weight Empty (lb)	9,548	16,015
Combat Weight (lb)	15,533	31,441
Max Takeoff Weight (lb)	24,500	33,500
Number on a CVA-19	173	130
Guns/Ammo	2/200	2/720
Flyaway Cost ($)	700,000	1,400,000
Internal Fuel (gal)	810	1,500
Combat Radius (nm)	195	650
Bombing Accuracy (CEP)		
Mils	30 to 50	10.3
Ft	150 to 250	134
External Stores Stations	5	6 wing
		2 fuselage
Engine		
Model	P&W J52-P-6	P&W TF30-P-6
Military Thrust (lb)	8,500	11,350

The report was as complimentary as these ever are. The automatic weapons-delivery system was as accurate as desired although the gunsight had a boresight retention problem. The Mk 12 guns were not reliable. Neither the radar nor the ECM/DECM were considered satisfactory. "Aircraft performance was found to be satisfactory although an increase in power available would enhance the mission potential." The exhaust was considered too visible, apparently leaving an even more prominent trail than the F-4. Minimization of visual signature had been elevated in importance based on combat operations in Vietnam. Ironically, one of the problems that plagued the aircraft operationally, hydraulic leaks, was not one of the deficiencies highlighted.

Although otherwise rated highly with respect to the basic carrier suitability requirements, there was an unintended consequence stemming from the combination of the low-mounted intake, the steam catapult, and the turbofan engine. As determined early on during shore-based trials, the engine tended to choke on the superheated air being ingested during the catapult stroke. After compressor modifications, initial at-sea carrier qualifications were accomplished from *America* in November 1966. This resulted in permitted thrust on a catapult launch being reduced from 10,560 to 8,810 pounds, which meant a maximum gross weight reduction to 34,000 pounds. However, subsequent catapult track modifications to minimize steam leakage and rescheduling of the engine's 12th stage bleed valve (incorporated in the -6A engine along with compressor changes), restored the maximum takeoff weight to 38,000 pounds.

The A-7 was initially qualified for Mks 28, 43, 57, and 61 nuclear weapons. According to the BIS report issued by the U.S. Naval Weapons Evaluation Facility at Kirkland AFB, New Mexico, "The low altitude navigation capability and the excellent range and endurance of

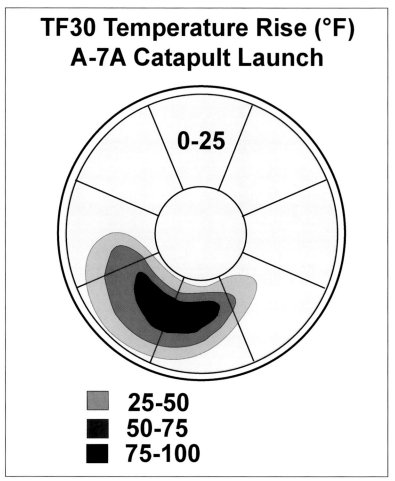

TF30 Temperature Rise (°F) A-7A Catapult Launch

0-25

■ 25-50
■ 50-75
■ 75-100

Jet engine performance decreases with increased ambient temperature, which is bad enough, but the temperature rise from steam ingestion was very localized, which made compressor modifications to accommodate the situation all the more difficult. (Author)

the A-7A airplane significantly enhance its nuclear weapons service suitability." One major shortcoming identified was the inability to change delivery parameters in the weapons computer while in flight.

Shrike, Bullpup, Walleye, and Sidewinder qualification was accomplished at the Naval Missile Center, Point Mugu, California, and the Naval Ordnance Test Station, China Lake, California, in late 1966 and early 1967. At least one Bullpup launch from an inboard station caused an engine compressor stall and flameout, although it automatically relit immediately. The inboard stores stations could also not be used with the speed brake extended.

VA-147 was the first to take the A-7A into combat, flying from *Ranger* in December 1967. During this first deployment, they flew 1,400 sorties with the loss of only one aircraft. Although the A-7 had better payload-range performance than the A-4 it was to replace, experienced jet attack pilots didn't immediately prefer it. Relative to the A-4E, it seemed underpowered, both in fact and because of the

The A-7 could carry something for everyone. In this publicity photo taken near the Naval Air Weapons Station at China Lake, California, in the late 1960s, this A-7 was loaded from outboard to inboard on each side with an anti-radar missile, a Walleye, a big Bullpup, and a Sidewinder. (Vought Aircraft Heritage Center)

acceleration characteristics of the turbofan engine compared to the axial flow engine. It was also considered to be less maneuverable, in that it had less margin from an accelerated stall and was prone to spin if mishandled. Hydraulic leaks were an annoyance, ironically caused by the use of F8U hardware.

The A-7 replaced not only A-4s on the carriers, but also A-1s in squadrons that did not transition to the A-6. The SPAD pilots weren't impressed by the A-7's endurance but were very happy to have more speed, a multi-mode radar, and air conditioning. The rolling map display would have been more appreciated if it had been more reliable.

A total of 193 A-7As were produced, including the three prototypes.

A-7B

The Navy had originally intended to incorporate its Integrated Light Attack Avionics System (ILAAS) in the A-7 beginning with the 200th production airplane. It was to be a digital system as opposed to the collection of existing avionics managed by an analog computer in the A-7A. It would also provide a heads-up display that presented all the necessary data to the pilot to perform an accurate attack while maneuvering in the dive.

Unfortunately, there were delays in the development of the ILAAS. However, the A-7B did have an uprated engine, the TF30-P-8 with 12,290 pounds of thrust[8]; a modification of the flap system to allow setting at any position[9]; improved brakes; an improved Shrike display; additional ECM; and an attempt to reduce hydraulic leaks by incorporating new fittings. The maximum speed-brake extension was limited to 50 degrees instead of 60 degrees as it had been in the A-7A because of excessive buffet at full extension. The empty weight only increased by about 300 pounds but with no corresponding increase in the gross weight of 38,000 pounds, so the useful load went down by the same amount.

The additional Electronic Countermeasures (ECM) included passive and deceptive capability: The ALQ-100 transmitted a false return for track breaking of threat radars. The ALE-29 was a chaff dispensing system. The APR avionics provided relative bearings to threat radars and specific warning of a SAM launch.

BIS trials were accomplished in mid-1968, including at-sea testing from *Lexington* in August, with a final report issued 12 December. Poor reliability of the automatic weapons-delivery system was still a problem. Even though the engine had been uprated in thrust, the BIS report noted: "The full mission potential of the A-7B aircraft remains unrealized due to the thrust limitations of the engine, as was the case with the A-7A." The engine was still susceptible to compressor stall during weapon firing and catapult launch. Launches had to be made from "super-sealed" catapults with a minimum of 20 knots of WOD to minimize steam ingestion and provide an acceptable compressor stall margin.

The A-7B BIS report noted: "There is little difference in the performance of the A-7A/A-7B airplanes. The fleet has voiced many

The first squadron to take the A-7A into combat was VA-147 deploying aboard Ranger in November 1967. Here one is about to be catapulted with a full war load of two Shrikes, four CBU-29/B cluster bomb containers, and two Sidewinders. The Shrikes were to disable the SAM tracking radar antenna and the cluster bombs, the personnel, and the rest of the equipment at the SAM site. (Vought Aircraft Heritage Center)

As part of demonstration and validation of the A-7A's takeoff performance, launches were accomplished with full internal fuel and four 300-gallon external tanks, a total fuel capacity of 17,000 pounds. In May 1967, as part of the Navy's participation in the Paris Air Show, two A-7As were used to demonstrate the ferry range in this configuration. CDR Charles Fritz, USN, and Capt. Alec Gillespie, USMC flew from NAS Patuxent River, Maryland, to Paris, France, non-stop and unrefueled. The flight took about seven hours, and the pilots landed with two hours of fuel remaining. (Vought Aircraft Heritage Center)

The A-7B was a minimal change to the A-7A. Part of the change was additional ECM capability, notable here as the radar warning antenna at the upper trailing edge of the vertical fin. Zuni rocket pods are loaded on the outboard pylon. These each contained four 5-inch-diameter folding-fin rockets. (Vought Aircraft Heritage Center)

A diving attack was facilitated by a large dive brake, similar to but even larger than the one on the F-8. Fully extended, it would limit the airspeed to about 400 KEAS at 10,000 feet in a 50-degree, idle-power dive from 35,000 feet. However, it could not be extended when releasing stores from the inboard wing stations because of the risk of the store yawing enough immediately after release to strike it. The external store on this VA-12 A-7E is a self-contained refueling pod. (U.S. Navy DN-SC-87-05648)

The USAF A-7D was a very successful adaptation of the Navy A-7A/B. It differed in detail in consideration of the Air Force's different equipment and requirements. For example, an in-flight refueling receptacle, visible here atop the fuselage at the leading edge to the wing, was substituted for the Navy's in-flight refueling probe. (Vought Aircraft Heritage Center)

complaints regarding the lack of performance of the A-7A airplane in coordinated air wing strikes. The A-7A airplane performance, particularly with loadings in excess of 100 drag counts, required that the entire strike group slow down to accommodate the A-7s.[10] This is very detrimental to the overall success of the mission. The slower the strike group approaches the target, the less will be the effect of the important element of surprise. Also the longer the strike group takes to travel to the target area the less fuel the fighter cover aircraft will have remaining to engage the enemy and return to base."

Combat experience also increased the emphasis on survivability. The A-7's flight-control system was similar to the F-8's. Although there were two separate and independent hydraulic systems for the flight controls, the control surfaces were positioned by tandem actuators and the hydraulic lines tended to be routed together. A hit on an actuator or hydraulic lines would potentially result in the loss of fluid in both systems, causing a loss of control. As a result, the actuators were now protected by armor plate and hydraulic lines were rerouted.

VA-25, the last operational AD-6/7 squadron, received 13 factory-new A-7Bs on 1 November 1968. After a relatively brief workup and training period, the squadron returned to combat aboard *Ticonderoga* in February 1969 for the first deployment of the A-7B. A total of 196 were produced.

U.S. Air Force Participation and Improvements

Although VA(L) was originally intended by OSD to be a joint-service program, the Air Force preferred to use existing fighters for close-air support. In the summer of 1964, however, it was encouraged by OSD to consider low-cost alternatives for this mission requirement, which it did in a study that was completed in December 1964. The alternatives considered included the F-5, which was a small supersonic Northrop fighter exported in the U.S. Military Assistance Program, and the A-7. The study concluded that the former was superior to the latter insofar as the Air Force defined the requirement.[11]

Secretary McNamara then requested a close-air support airplane procurement plan from the Air Force, noting that dwindling levels of former Navy A-1 Skyraiders necessitated an IOC in 1967. This limited options to an existing production aircraft. The F-5, A-6, and A-7 were mentioned as potential candidates. OSD's Systems Analysis Group was separately lobbying with the Air Force's Tactical Air Command (TAC) for the A-7. TAC preferred the F-4, even if its air-to-air capability was diminished by stripping out the Sparrow missile system. The superiority of the F-4 was subsequently confirmed in a TAC study that projected a higher loss rate to enemy air and ground-based defenses of the A-7 due to its slower speed and that it was significantly inferior overall. Nevertheless, OSD continued to press for a low-cost alternative. The war in Vietnam was increasingly demanding something very much like the Skyraider in terms of payload and endurance, which the F-5 definitely wasn't, as it subsequently demonstrated in an in-country evaluation that began in October 1965.

In November 1965, the Air Force finally decided to buy the A-7, but only after addressing its shortcomings, starting with engine thrust.

Since it was a Navy program, the Air Force equivalent of a Program Manager was assigned to the Navy A-7 Program Management Office at the Bureau of Weapons in Washington, D.C.

The Navy and Vought had already considered increasing the thrust of the TF30-P-6 in the A-7A by adding an afterburner. It had not been as pressing an issue to the Navy as it was to the Air Force because of the takeoff assist of the steam catapult. In January 1966, however, the Navy requested permission to add an afterburner to its A-7s. OSD denied the request, reportedly because an afterburner would hurt range with no commensurate increase in mission performance or survivability.

The Air Force also wanted a more advanced avionics system for all-weather operation and detailed changes, like the substitution of its refueling receptacle for the probe system used by the Navy, the addition of anti-skid brakes (the Navy A-7 was hard to stop on a wet runway), and incorporating its 20mm M61 Gatling-type cannon in lieu of the two Navy 20mm Colts.[12]

The Air Force request to add an afterburner to its A-7 was disapproved as the Navy's had been. Plan B was to incorporate the Rolls-Royce Spey built under license by the Allison Engine Company as the TF41. The Spey was a proven engine, flying in airliners as well as the Royal Navy's Buccaneer and selected, with afterburning, for the British F-4 Phantoms. It would provide a thrust increase of almost 3,000 pounds and slightly lower specific fuel consumption. OSD concurred with the Air Force recommendation in August 1966. The Air Force awarded a contract to Allison in December.

Meanwhile, the Navy and Air Force A-7 deputy program managers had teamed up to jointly promote an avionics upgrade program, including elements from the ILAAS program, to improve the weapon system's accuracy in the combat environment. LTV management was initially not supportive, since they were worried that a sophisticated avionics system would make the airplane unaffordable compared to buying more A-4s, which the Navy was in the process of doing to replace airplanes shot

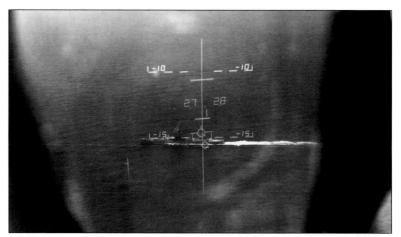

A-7D/Es were among the first operational airplanes with a heads-up display, which replaced the gunsight and also provided essential flight data like airspeed, attitude, and altitude. (Vought Aircraft Heritage Center)

down in Vietnam.[13] The government program managers finally got LTV's attention by threatening to place the contract with an avionics company instead of Vought.

The new avionics system, which OSD approved, was to be all-digital with advanced flight and navigation displays, a general purpose digital computer, an inertial navigation system, and a heads-up display. The computer not only integrated the data from all the inputs available, but it could also reject an input it judged erroneous and substitute data from another source. Data was projected on the heads-up display (HUD) as required for navigation, terrain following, weapon delivery, and landing approach.

The real-time computation of weapon delivery, Continuously Computed Impact Point (CCIP), meant that the pilot need not be at a particular dive angle, airspeed, altitude, or g to achieve bombing accuracy. The Bomb Fall Line (BFL) was shown on the HUD with an "X" displayed on the line where the ordnance selected would hit if released at that instant. The pilot simply had to steer the X over the target and drop the bomb. If the target was moving, the pilot needed to lead it appropriately. Alternatively, the pilot could designate the target with an aiming reticle, commit the system to drop, and then follow steering commands on the heads-up display. The Continuously Computed Release Point (CCRP) mode would automatically drop the selected ordnance when the proper release point was reached.

The 459 USAF A-7Ds differed not only in the engine, gun, and avionics system to the A-7A/B, but in many other details because of operational differences with the Navy. These included a self-contained engine start capability, an in-flight refueling receptacle in lieu of a refueling probe, and the deletion of the nose gear launch bar.

A-7E

Although the Navy had recognized the need for more thrust in the A-7, it was not initially enthusiastic about changing to the TF41, in part because of a previous unhappy experience with Allison during the reengining of the F3H Demon with the J71. If nothing else, NavAir (formerly a part of BuWeps) feared another round of development problems with a new engine in the steam catapult environment. The Navy was also hopeful that Pratt & Whitney would come up with a more powerful version of the TF30. The increasing weight of the A-7 and the Air Force commitment to the TF41 overcame the Navy's reluctance and Vought's. However, the first 67 A-7s with the new digital avionics suite had to be delivered with the A-7A/B's TF30 engine since TF41s were not yet available. These were designated A-7C.

OSD resolved the impasse between the Navy and the Air Force over the selection of the gun for the A-7D by directing that both services buy new airplanes with the Air Force gun. It was a six-barrel rotating cannon that fired at a rate of 4,000 or 6,000 shots per minute. A full ammunition load was 1,018 rounds. Other A-7D features incorporated in the A-7E were the third hydraulic system and anti-skid brakes. Unique to the A-7E was a strike camera.

Vought reluctantly guaranteed an A-7E accuracy of 10 mils, which would result in a CEP of 140 feet with a bomb release at 10,000

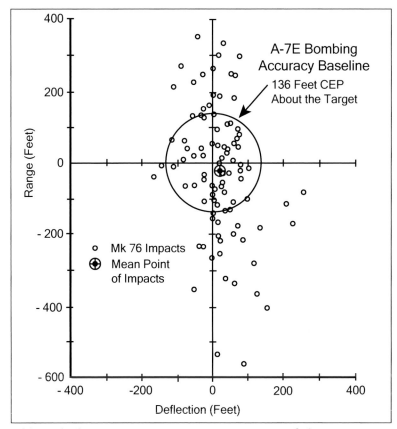

Although the A-7E weapon system was one of the most accurate fielded up to that time, the accuracy was relative. The Circular Error Probable (CEP) is the radius of a circle in which 50 percent of the bombs hit. This figure from the A-7E service acceptance trials report shows the CEP about the target and where the other 35 practice bombs landed. (All were dropped one at a time.) Note that the dispersion along the bombing axis, 900 feet, is more than twice that across it. For this chart, the test team corrected the impact points for those bombs dropped using a practice multiple bomb rack because the A-7E's computer was programmed for a 200-millisecond delay in weapon release instead of the measured delay time of 50 milliseconds. For reference, the average human's reaction time to an anticipated action is about 200 milliseconds. (Author)

feet. The A-7A/B was demonstrating accuracies of less than 20 mils, with 10.3 measured during A-7A service acceptance trials. As it turned out, when the new avionics system was working, A-7E pilots were able to achieve even better CEPs. The A-7E Armament Service Acceptance Trials report dated 27 February 1970 stated that it had the potential to deliver low-drag bombs with a CEP of 4 to 6 mils. Whereas CEPs with a conventional gunsight in the A-7A/B were 100 to 150 feet in training and often twice that in combat, the A-7D/E system resulted in CEPs of on the order of 50 feet in training and, anecdotally, less than 100 feet in combat. Moreover, that level of accuracy

Red shirts denoted ordnance men aboard the aircraft carrier, here loading 20mm ammunition into and spent cartridges out of the A-7E's ammunition drum, which was located in the fuselage directly aft of the canopy. (Vought Aircraft Heritage Center)

The A-7E cockpit included a much-improved moving map display that could be directly compared with the radar presentation. The new features and functions like the large heads-up display provided a significant improvement in workload reduction, situational awareness, and all-weather mission capability for the single-seat attack pilot. (Vought Aircraft Heritage Center)

The A-7E's precision weapon delivery wasn't always used. Below, VA-97 Corsair IIs are dropping on command from an Air Force Pave Phantom F-4D, which was Loran-equipped for what was considered precise position determination in pre-GPS times. (Vought Aircraft Heritage Center)

The final major update to the A-7 was the addition of an external FLIR pod, which provided increased night and all-weather capability and accuracy compared to using the radar for weapons delivery. The A-7's FLIR pod was entirely self-contained. First delivery of a new production A-7E with the FLIR pod was September 1978. VA-81 and VA-146 were the first East Coast and West Coast squadrons, respectively, to deploy with the FLIR. (Vought Aircraft Heritage Center)

Mines were often employed to disrupt and control enemy shipping. The A-7E was particularly useful for planting mines because of its accurate navigation system, since sooner or later the Navy would be obligated to clear the mines. Knowing where they were dropped is helpful in this regard. This A-7 is carrying an Mk 52 mine. (Vought Aircraft Heritage Center)

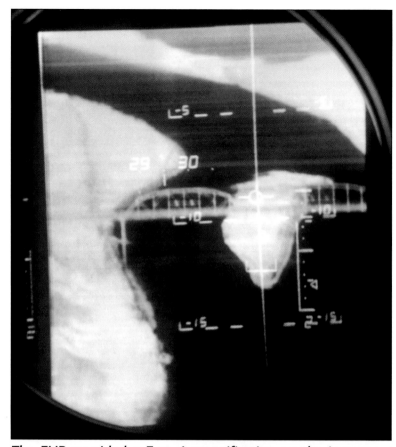

The FLIR provided a 7 to 1 magnification on the instrument panel display shared with the radar and had a field of view of 40 degrees, 5 degrees upward and 35 degrees downward. The FLIR image could also be displayed on the heads-up display. It was particularly useful against high temperature-contrast targets like a ship at sea or as shown here, a bridge. (Vought Aircraft Heritage Center)

resulted in mission capability that "can easily deteriorate to a point where it is inferior to its predecessor, the A-7A/B" according to the Service Acceptance Trials report. It was also initially not approved for delivery of nuclear weapons except in a VFR manual laydown mode due to errors in the data from the new APQ-126 radar and the APN-141 radar altimeter. According to George Spangenberg:

> We at the beginning were using all off-the-shelf items because we had to and we had a semi-all weather kind of updated system under development but it was not as far along as what the Air Force put into their A-7D. Later of course we picked up much of that same avionics, went to the fleet as the A-7E. It was almost rejected by the fleet as altogether too complex. It went through a period of about six months with Op Eval saying that degree of complexity could not be handled and the then program manager Shepherd doing all kinds of studies on "what do we do," "how do we get out of" the troubles. Eventually the system got worked out and the fleet learned to live with it. And you had to have that capability. We would have been extremely shortsighted if we hadn't gone ahead with it so the Air Force introduction into the program helped the Navy capability in the long run. And obviously the Air Force got a capability that they didn't have anywhere else. It was good for the services but these bi-service and tri-service programs really only work when you start with the Navy designed airframe.

The A-7E was first deployed on *America* in May 1970, flown by VA-146 and VA-147.

The A-7E possessed a significantly greater mission capability than previous A-7s because of its more capable avionics system and somewhat better combat maneuverability. However, because fuel burn for the takeoff, climbs, and combat portions of the mission profile was computed at military/intermediate (30-minute rating) or maximum continuous/normal thrust settings, *not* thrust required, the TF41-powered A-7E's combat radius was significantly less than that of the TF30-powered A-7s.

	A-7A	A-7B	A-7C	A-7E
Engine	TF30-P-6	TF30-P-8	TF30-P-8	TF41-A-2
Basic Weight	15,982	16,477	18,250	19,575
Combat Weight (lb) 6,000 lb Fuel and 12 Mk 81s	27,736	27,657	29,385	30,745
Military Thrust (lb)	11,350	12,200	12,200	15,000
Weight/Thrust	2.44	2.26	2.41	2.05
Combat Radius (nm)	570	613	567	432

Maximum A-7E launch weight was 42,000 pounds, which provided 20,000 pounds of useful load. However, launch at this weight was limited to ambient temperatures of less than 50 degrees F. At 80 degrees F, the launch weight was limited to 38,000 pounds, the same as the A-7A/B/C. The A-7Cs actually had a slightly higher launch weight at temperatures above 80 degrees F.

was achieved in far fewer training flights. However, since a 500-pound bomb still had to hit within 25 feet of a truck to destroy it, multiple bombs still had to be dropped in an attack to insure destruction of a small target.

The A-7E deliveries began in July 1969. BIS trials were completed in the fourth quarter of 1969, with at-sea carrier-suitability testing accomplished from 24 to 26 October aboard *Independence* and on 24 November aboard *America*. Although the TF41 provided more thrust for takeoff and better engine acceleration characteristics on approach, the A-7 was now even heavier with the added equipment. The TF41 was not susceptible to compressor stalls from steam ingestion at launch, so the super-sealing catapult track and 20-knot WOD were no longer required.

The A-7E got off to a rocky start with the integrated avionics system not well received at first. When fully operational, it significantly reduced pilot workload and provided increased mission capability. However, its reliability was poor and failures of subsystems quickly

The empty weight of the A-7E with six wing pylons, 1,000 rounds of ammunition, four empty MERs, and two Sidewinder missiles was approximately 23,000 pounds. That left only 2,300 pounds of fuel at the maximum arrested landing weight. While satisfactory for VFR recoveries, it was considered marginally acceptable for night/IFR recoveries. Any hung ordnance would reduce the maximum fuel onboard for approach accordingly and to an unacceptable level.

The A-7E went on to replace all the A-7As, Bs, and Cs with 535 built. One late modification, circa 1980, was the addition of automatic maneuvering flaps. These reduced pilot workload and the propensity of the A-7 to depart when heavy and pulling gs. Toward the end of its operational use, squadrons would remove one of the wing station pylons from each side to reduce drag.

Maverick

The Air Force developed the AGM-65 Maverick air-to-ground missile with a 125-pound warhead in the late 1960s. Guidance was originally by TV. The pilot used the seeker in the missile to select and lock on a target. When fired, the missile would then guide to that feature while the launching aircraft departed the premises. It was initially fielded in 1972.

The laser Maverick and the IR Maverick were identical except for the seeker heads. (Vought Aircraft Heritage Center)

The A-4F was the last single-seat Skyhawk model procured by the Navy for carrier-based attack squadrons. Many of the last of these, the Super Foxes as shown here, were assigned to the three VA squadrons that deployed on Hancock. This one, blocking the view of an A-7A which was supposed to have replaced the Skyhawk, has the slightly bulged engine inlets for the Super Fox's J52-P-408 engine and the fairly rare, for an A-4F, ALR-45 radar warning antennae on the nose and fin tip. Some Hancock A-4Fs were equipped with laser detection for dropping laser-guided bombs designated by the rear seater in a TA-4F. (Robert L. Lawson collection)

The A-7 was spurned by the Marine Corps in favor of production of yet another significant upgrade to the A-4. Only the Marines operated the A-4M in squadron strength. This one was assigned to VMA 223 and photographed at NAS Corpus Christi in April 1984. The Israelis bought the similar A-4N for their air force. (Philip Fridell)

The Navy adapted the Maverick in the early 1980s, substituting infrared image homing and a significantly bigger 300-pound warhead to be used primarily as an antiship weapon. The AGM-65F IR Maverick was fired from an A-7E at a destroyer target in 1983, but it wasn't deployed until 1989. (A laser-guided AGM-65E was developed for the Marine Corps and first fielded in 1985.)

A-4s Nearly Forever

Although Douglas lost the 1963 competition, it was not the end of the line for the Skyhawk. For one thing, some pilots preferred the Skyhawk to the Corsair II, primarily for its maneuverability and better thrust-to-weight ratio, which had been sacrificed for maximum fuel and bomb-carrying capacity. The A-4 was arguably more combat damage-tolerant since it had a manual backup for the flight controls rather than relying on a second hydraulic system as the A-7 pilot did. The Skyhawk supporters did acknowledge that the A-7 was superior in range, payload, loiter time, and some mission avionics. However, the A-7 cost more to buy, so much so that Douglas could propose fairly expensive upgrades and remain competitive.

Douglas production of the A-4E (A4D-5) ended in 1966, slightly overlapping the A-7 production startup. However, because of attrition in the Vietnam War and to insure a production base if the A-7 was delayed or disappointed, Douglas had received a contract in 1965 for delivery of 207 A-4Fs beginning in 1967. (The last 60 of these were canceled when the A-7 had demonstrated its worth.) The A-4F addressed the growing volume of avionics required by new offensive and defensive capability with the addition of a large removable hump immediately behind the cockpit extending aft to the dorsal fin. It also had nose-wheel steering, large wing-lift spoilers, a zero-zero ejection seat, and the uprated P&W J52-P-8A engine with 9,300 pounds of thrust. After delivery, many of the A-4Fs were retrofitted with the J52-P-408 with 11,200 pounds of thrust and slightly bigger engine inlets because of the higher mass flow. These were familiarly and fondly referred to as the Super Fox. The Blue Angels, the Navy's flight demonstration team, began flying the Super Fox, stripped of the hump, in late 1973 with the first shows in 1974. The A-4 also became an "adversary" aircraft for fighter pilot tactics training. Its small size and maneuverability made it ideal as a MiG substitute and it humbled many a budding fighter pilot while providing him the necessary experience to survive real-world combat.

In 1969, the Marines, who were expected to replace their A-4s with A-7s, opted for 160 A-4Ms instead. Range was not nearly as significant to the shore-based Marines as it was to the Navy and unit cost was. The A-4M incorporated a significant avionics upgrade to the A-4F with

The diminutive A-4 Skyhawk was the first non-fighter to be used by the Blue Angels (the request for F-14s and then A-7s supposedly being turned down). It replaced the F-4, which was expensive to operate and accident-prone. Even without afterburner, it was a good airshow airplane, having an excellent roll rate. (U.S. Navy DN-SC-84-09556)

Official Navy restrictions on squadron paint schemes and markings varied over the years. They were particularly onerous after tactical paint schemes were introduced, since any showmanship defeated the purpose. However, for moral purposes, permission was sometimes granted for one aircraft in each squadron in an air wing to be more colorful than regulation. These special markings were usually applied to the CAG's designated airplane, in this case a post conflict commemoration of VA-72's participation in Operation Desert Storm. Note the mission markers, camels, between the pilot's name and the modex. (Vought Aircraft Heritage Center)

In the late 1960s, at the behest of the Army, the Air Force finally broke with its practice of using fighters for close air support and initiated a program to develop an airplane designed specifically for the mission, with emphasis on destroying tanks and other armored vehicles. The result, after a fly-off competition in 1972, was the Fairchild-Republic A-10 Thunderbolt II, more familiarly known as the Warthog. Part of its armament was a huge 30mm cannon. There were also 11 stores pylons, which could be loaded to the hilt as shown here. The A-10 won a fly-off against the A-7D in 1974 that resulted in the gradual relegation of the Air Force Corsair IIs to Air National Guard units. (National Archives)

other configuration changes as well, such as an enlarged canopy, the J52-P-408 engine, and twice the number of rounds per gun. To allow for operation from expeditionary airfields, a self-contained engine start capability was provided, the tailpipe was canted slightly upward to allow for quicker liftoffs, and a drag chute was added.[14] Production of the A-4 concluded in February 1979 with the 2,960th also being the last A-4M. Although carrier qualified, it was never operated in strength by Navy squadrons. The Marines retired the last one from a reserve squadron in 1994.

A-4s proliferated throughout the world, mostly in land-based air forces. Most were surplus U.S. Navy airplanes, but the Israelis, who also evaluated the A-7, opted for new production A-4Ns based on the A-4M airframe but incorporating 30mm cannons in place of the 20mm, yet another avionics system update, and an IR suppressor. Cost was probably a factor, but, as with the Marines, range was not a major criteria and all-weather capability was not required. The 117 A-4Ns were produced between 1972 and 1976.

The A-4N procurement sparked an exchange of letters to *Aviation Week* in 1972 by U.S. Navy A-4 and A-7 pilots. The first appeared in February and/or March 1972. One was from a Navy commander named Jerry O. Tuttle, who went on to become a VADM. However, in the very next issue, a lieutenant commander, whose name was withheld by request, advocated that "Congress and the Navy

should investigate procuring the (A-4N) for use by both fleet and reserve attack squadrons. The unit cost of the A-4N should be considerably less than for the A-7E."[15] *Aviation Week* subsequently printed at least two rebuttals, one from a Navy A-7 pilot.

The A-7 outlasted the A-4 from a production standpoint by only four years, the last A-7E being delivered in 1983. (Upgrades and modifications continued for a few years after that.) Vought also did not do nearly as well at selling more A-7s as Douglas did while its successor was being fielded. Nevertheless, the Corsair II was a very successful program and one of only three combat airplanes up until then that was operated in strength by both the U.S. Air Force and the U.S. Navy. The A-7E's last deployment was with VA-46 and VA-72 aboard *John F. Kennedy* as part of Operations Desert Shield and Storm. Their contribution included the predawn strike on radar sites in and around Baghdad on 17 January 1991 and continued until that deployment of "Big John" ended in March.

Ironically, the A-7s in the Air Force and the Navy would be replaced by airplanes which couldn't have been more different. The Air Force, which had dropped the attack designation in 1946, and before the A-7 had used supersonic fighters for close air support and interdiction missions, procured the decidedly subsonic A-10. The Navy, which had adopted the attack designation in 1946 and continued the separation between its fighter and attack communities for another 30 years, procured a supersonic strike fighter and assigned it to dual-role VFA squadrons.

CHAPTER TEN

ONE IF BY LAND,
TWO IF BY SEA

An F/A-18C carrying two bombs, one laser-guided and two Sidewinders, is launched from Abraham Lincoln *in late 2002 to go on patrol over Iraq. As configured, this Hornet can accomplish both air-to-air and air-to-ground missions on demand.* (U.S. Navy 020921-N-9593M-086)

In the early 1970s, OSD pressed the Navy to develop a low-cost fighter. The F-14's affordability had worsened in 1972 when Grumman realized that its production contract with fixed price options did not adequately incorporate inflation and demanded a price increase to continue. Congress was limiting F-14 production to only 334 aircraft at a rate of 50 per year, which meant that all Navy and Marine fighter squadrons could not be equipped with F-14s, necessitating F-4 service life extensions. After considering alternatives including a navalized F-15, the Navy conducted the VFAX fighter study that resulted in the recommendation to replace both the F-4 and the A-7 with a common airplane. As a fighter, it would augment the F-14s, allowing a reduced number of Tomcats to be acceptable.

The Navy wanted the survivability, maintainability/reliability, and cost of the VFAX design to have equal emphasis with performance.[1] Maximum speed required was only Mach 1.6. Acceleration was more important, a requirement perhaps influenced by the relatively sluggish

When Vought marketing arranged this lineup, they must have felt pretty good about their chances of winning the Navy's new VFAX program. The F-8 was considered by many to be the last of the gun fighters, the A-7 had an excellent reputation as an attack aircraft, and Vought had teamed with General Dynamics to propose a variant of the F-16 to the Navy that combined both fighter and attack requirements. (Vought Aircraft Historical Foundation)

performance of the otherwise competent Corsair II. One desire was that the basic configuration be adaptable to a V/STOL derivative to serve as a replacement for the Marine Corps AV-8s and A-4s.

OSD approved the plan, and the Navy issued a pre-solicitation notice in June 1974. Seven contractors responded with predesign studies by mid-July. In the meantime, the Air Force had reluctantly responded to similar pressures with its Light Weight Fighter (LWF) program. The result was two-airplane demonstrator contracts in April 1972 to General Dynamics and Northrop. These were designated the YF-16 and the YF-17, respectively, and to be small, minimally equipped dogfighters, which could be bought in quantity because of their low unit price. Armament was limited to a 20mm M61 cannon and two Sidewinder missiles. The YF-16 flew in January 1974, albeit inadvertently the first time, and the YF-17, in June. Both General Dynamics Ft. Worth and Northrop responded to the VFAX pre-solicitation studies with derivatives of their LWFs.

In September 1974, Congress unexpectedly put the kibosh on a separate Navy new fighter program and directed that the Navy buy a derivative of the Air Force's LWF. (The Air Force LWF program was also known as Air Combat Fighter, which is why the Congress-dictated program was the Navy Air Combat Fighter.) The following language was included in the Joint Committee on Appropriations Report:

The (House and Senate conferees) are in agreement on the appropriation of $20,000,000 as proposed by the Senate instead of no funding as proposed by the House for the VFAX aircraft. The conferees support the need for a lower cost alternative fighter to complement the F-14A and replace F-4 and A-7 aircraft; however, the conferees direct that the development of this aircraft make maximum use of the Air Force Light Weight Fighter and Air Combat Fighter technology and hardware. The $20,000,000 provided is to be placed in a new program element titled "Navy Air Combat Fighter" rather than VFAX. Adaptation of the selected Air Force Air Combat Fighter to be capable of carrier operations is the prerequisite for use of the funds provided. Funds may be released to a contractor for the purpose of designing the modifications required for Navy use. Future funding is to be contingent upon the *capability of the Navy to produce a derivative of the selected Air Force Air Combat Fighter design.*

NavAir was forced to cancel its planned procurement and redirect it to the two LWF contractors. Since neither of them was a traditional Navy contractor experienced with carrier-based airplanes, each teamed with one that was, General Dynamics with Vought and Northrop with McDonnell.

The major difference between the two LWF fighters was that the YF-16 had one engine and the YF-17 had two. Both designs used fly-by-wire control systems, but the YF-17 had a mechanical backup. General Dynamics installed the ejection seat at a more reclined angle than usual under a single-piece bubble canopy, the former for lower frontal area[2] and the latter for better visibility. Northrop's design incorporated a wing leading edge root extension (LERX) bigger than the one on the YF-16 and also twin vertical fins. The combination permitted controlled flight at extremely low speeds and high angles of attack.

After a relatively brief envelope expansion by the contractors, the Air Force began to conduct a fly-off in parallel with each company's ongoing development. Each airplane was superior to the other in some respects and neither was inferior to the requirement. In January 1975, the secretary of the Air Force announced that the General Dynamics YF-16 had been selected for full-scale development. If there was a significant tiebreaker, it was probably that the single-engine YF-16 was

less expensive, but its better range and transient maneuverability were also notable. It was powered by the Pratt & Whitney F100 engine like the Air Force's F-15, whereas the YF-17's engine, the General Electric J101, had just been qualified and had not been fielded. It also had a lower bypass ratio than the F100, which meant it was less fuel efficient.

In the meantime, the new team member for each airplane was refining its configuration to meet the Navy's VFAX requirements. Not only was it to be carrier compatible—capable of catapult launch and arrested landing, provided with folding wings, etc.—but it was also to have more internal fuel, more air-to-ground capability, a bigger radar, and Sparrow missile armament. Proposals were submitted to the Navy in December 1974. Because of the different mission and operational requirements, including addition of radar-guided missiles, both airplanes were significantly different from the prototypes in detail.

Vought initially proposed two variants of the YF-16, its Models 1600 and 1601. The 1600 was powered by the F-14's intended Pratt & Whitney F401 engine and fully responsive to the Navy's mission requirements. It was more of a departure from the Air Force F-16 configuration than the 1601, which was powered by the F-16's F100 engine and incorporated the F-16's avionics and armament. The empty weight was 18,454 and 16,876 pounds, respectively. The 1601, because of its commonality with the F-16 and lesser air-to-air capability, was less expensive. Both still required significant changes to the wing to accommodate the higher gross weight and slower approach speeds

In preparing for the VFAX competition, McDonnell was also on familiar and comfortable ground with a twin-engine airplane, Northrop's YF-17, and the Navy as a customer. (National Archives)

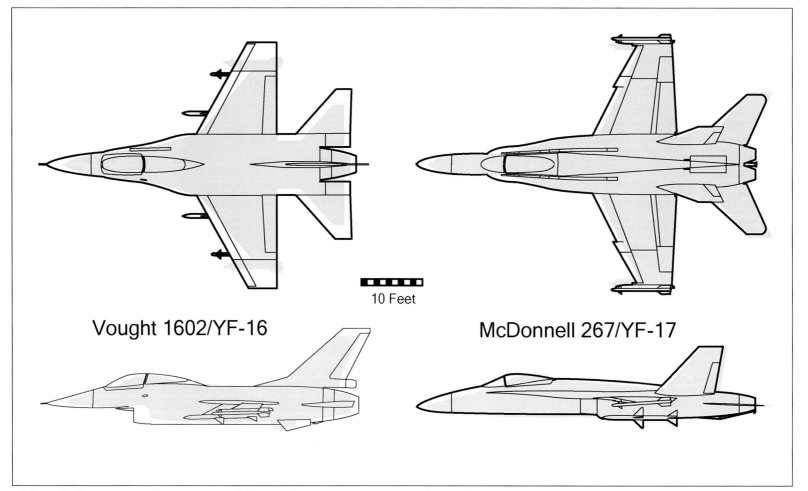

Vought 1602/YF-16

McDonnell 267/YF-17

10 Feet

Although the McDonnell 267 was as different from its YF-17 Air Combat Fighter baseline as the Vought 1602 was from the YF-16, at first glance it looked like somewhat less of a stretch. (Author)

dictated by the Navy specification. The area was increased and aileron effectiveness at slow speeds provided by an increase in travel and the addition of boundary layer control. A full-span Krueger flap was added to the leading-edge flap for maximum wing area and camber with the flaps down. Many other operational/service specific systems like the landing gear and in-flight refueling were also different as they were with the A-7D versus E.

Like Vought, McDonnell retained the aerodynamic configuration of the Northrop YF-17 but did an extensive redesign of the details and scaled the size and weight upward to meet the Navy's operational and mission requirements. For example, the wing area was increased from 350 ft^2 to 400. A snag was added to the wing leading edge to provide an aerodynamic fence to minimize spanwise flow at approach angles of attack. The leading edge and trailing edge flap deflection was increased by 50 percent and the ailerons drooped when the flaps were extended. The engineers retained the YF-17's mechanical backup for the elevator control system that provided emergency pitch and roll control, in part to alleviate any residual concerns in NavAir about fly-

by-wire. The cockpit was virtually all new, with the use of three large multi-function displays and many control functions added to the throttles and control stick.

In early January, the Navy had rejected the proposals from both companies, as neither airplane was considered fully responsive. In the meantime, the Air Force had announced its decision. Given the Congressional language, it would seem that the Navy had no choice but to buy the Vought airplane at that point. Instead, it proceeded as if it were free to make an independent selection. Surprisingly, OSD concurred. As a result, McDonnell elected to continue to compete, big production program opportunities being few and far between.

The major change to the revised McDonnell proposal was the substitution of an uprated version of the J101 engine, the F404, with a higher bypass ratio and an increase in thrust of 10 percent at sea level. This improved range, acceleration, and maximum speed and ceiling.

The revised Vought proposal, the Model 1602B submitted on 4 March 1975, increased mission performance and suitability, but also increased the divergence from the F-16. What was common—basic

The original Navy plan was to have McDonnell produce two slightly different airplanes, the F-18, shown here with Sparrows and Sidewinders, and the A-18, shown here with four laser-guided bombs as well as Sidewinders. (Jay Miller collection)

	Require-ment	Vought 1602*	McDonnell 267
Fighter Radius (nm)	400-450	400	415
Strike Radius (nm)	550	685	655
Maximum Mach (Intermediate Thrust @ 10,000 ft)	0.98-1.0	0.94**	0.99
Combat Ceiling (ft)	45-50,000	42,650**	49,300
Specific Excess Power (ft/sec)	750-850	723**	756
Acceleration from Mach 0.8 to 1.6 (sec)	110-80	105	88
Sustained Buffet-Free Load Factor	7.0-5.5	5.26**	6.6
Minimum Carrier Approach Speed	125-115	125	130**
Single Engine ROC (ft/min)	500	N/A	565

* Because Vought had not withdrawn its original 1600 and 1601 proposals, the Navy included that data as well in its presentation.
** Unacceptable

In its cost evaluation, the Navy concluded that while the McDonnell/Northrop program was the more expensive of the two proposals, it did not consider the premium to be significant compared to the overall superiority of the F and A-18 for the Navy's requirement.

Selection Protest

LTV Aerospace management immediately filed a protest with the General Accounting Office (GAO).[3] They primarily relied on the Congressional stipulation that the Navy was to select the derivative of the Air Combat Fighter chosen by the Air Force since its design was not superior in performance. They also pointed out that in following that mandate and the Navy's Request for Proposal, they had proposed a design that had maximum commonality with the F-16, whereas their competitor was "permitted to propose an undeveloped engine which transformed a previously unacceptable airplane into the F-18 subsequently selected by the Navy."[4] The protest requested that the F and A-18 selection be set aside and the competition be "reopened to the entire aerospace industry under identical ground rules uniformly applied to all competitors."[5]

One of several other contentions in Vought's protest was that the Navy evaluated its development and production cost proposal as being low and revised it upward, significantly. Vought noted, "These inflated figures are inexplicable and unrealistic in view of the company's track record. During the past 27 years, LTV Aerospace has come within an average 2.2 percent of target costs in its aircraft and missiles programs."[6]

Vought also took umbrage at, and prepared a detailed technical rebuttal of, the Navy's evaluation that the Model 1602 was deficient in carrier compatibility. According to the Navy, the airplane could attain a high enough pitch attitude in a carrier landing to experience "tail bumping." The Navy's design specification stated that the "No damage

shape, flight control, and electrical systems—and somewhat common—hydraulic and environmental control systems—was a far shorter list than what was not. For example, the wing aspect ratio and area was increased, the wing sweep reduced, and the wing camber eliminated in addition to the leading- and trailing-edge flap changes that had already been proposed. The airframe structure was similar in layout but considerably different in detail. If the intent of Congress was that the Air Force and Navy buy a very similar small fighter, the Navy operational and mission requirements had resulted in the F-16A and the Vought 1602B being very different other than in general appearance, just like the Northrop YF-17 and McDonnell 267 were.

In May 1975, the Navy announced that it had selected the McDonnell/Northrop YF-17 look-alike and designated it the F-18 for its fighter requirement and A-18 for the attack variant. Some of the Navy's stated and unstated reasons for preferring the YF-17 derivative were:

- Two engines versus one
- More tailored to the Navy requirement
- Better carrier compatibility
- Mechanical backup for fly-by-wire
- Not tied to a USAF program

According to its subsequent presentations to Congress, the two airplanes compared to its requirements as follows:

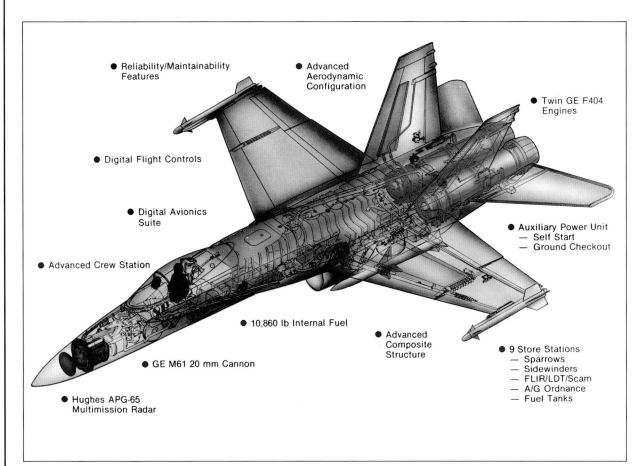

The F/A-18 was crammed full of capability, advanced technology, and design features. The location of the cannon immediately above and behind the radar was a high-risk decision, but it proved to be an acceptable juxtaposition of electronics and explosions. Note the fuel tank in the vertical fin, an indication of the maximum use of internal volume to meet the Navy's requirements while minimizing the increase in size over the YF-17. (Jay Miller collection)

ground line" angle equal or exceed the angle of attack at which 9/10 of the maximum coefficient of lift was achieved. The Vought proposal did not meet this requirement. However, Vought argued that (1) the F-16 angle-of-attack limiter precluded a pitch attitude that would result in a tail strike, and (2) the Model 1600 tail clearance at minimum approach speed was better than all but the A-4 Skyhawk of existing carrier-based aircraft, none of which had a tail bump problem in service.

The Navy's response to the issue of Congressional direction to buy the Air Force fighter was that it was not legally binding. It also addressed Vought's other complaints, but not to LTV's satisfaction. The company filed a detailed rebuttal with the GAO of the Navy's position that "it was free to ignore an expressed intent of Congress in choosing the F-18 aircraft."[7] Within a week, NavAir sent its detailed response to LTV's rebuttal to the GAO, opening with:

By letter dated July 14, 1975 LTV submitted its comments on the Navy's June 16, 1975 report on subject protest. We are certain it will be unnecessary to guide your Office page-by-page through the misrepresentations of the Navy's position contained in that document. Our June 16 report sets forth our views, and we only ask that it be referred to in the original rather than as misshapen in LTV's reply.

LTV disagreed with the Navy's projection of the time it would take to develop the "new" engine for the F and A-18 (a critical point in LTV's argument that a new competition would not adversely affect the availability of a new aircraft) and the Navy's technical and cost evaluation of LTV's proposal. They questioned the suitability of the F-18 for the attack mission, laying the groundwork for a comparison with the A-7. In a 21 July letter to the GAO, LTV's Washington-based counsel pointed out:

Further, it is noteworthy that even the Navy's evaluation of the F-18 concludes that the F-18 is marginal in the area of carrier suitability and carrier suitability/performance; marginal/ acceptable in the area of avionics; marginal in the area of ILS/Reliability/Maintainability; and unacceptable/marginal in the area of survivability. "Marginal" is defined in the report as a deficiency "to the extent that requirements can only be met with major change." "Unacceptable" is defined as deficiency "to the extent that major rework of the proposal, extensive negotiations, and re-evaluation would be necessary."

McDonnell faced another difficulty in that there was a Defense Department analysis that reviewed Navy estimates and concluded that

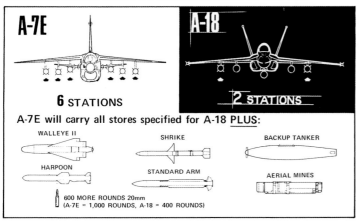

Ordnance Flexibility

STATIONS AVAILABLE FOR EQUAL RADIUS OF ACTION

A-7E — **6 STATIONS**

A-18 — **2 STATIONS**

A-7E will carry all stores specified for A-18 **PLUS**:

WALLEYE II

SHRIKE

BACKUP TANKER

HARPOON

STANDARD ARM

AERIAL MINES

600 MORE ROUNDS 20mm
(A-7E = 1,000 ROUNDS, A-18 = 400 ROUNDS)

Having lost the head-on competition and the protest, Vought then proceeded to a marketing campaign comparing the features of the "A-18" versus the A-7E, playing to its strength and range. In this page from a Vought brochure, the "A-18" is depicted as having to carry three external fuel tanks to match the radius of the A-7's radius of action on internal fuel and deficiency in weapons capability. The F/A-18 was eventually qualified to carry the Harpoon and the "Shrike" (what is depicted is actually the later AGM-88 HARM). It was never utilized as a tanker, however, because, unlike the A-7, it barely had enough fuel for its own purposes. (Vought Aircraft Heritage Center)

500 more F-14s could be bought for the $6-billion investment required to develop and produce the same number of F and A-18s. The offset was that F-18s would cost about $150 million less per year to operate than the F-14s. The White House's Office of Management and Budget was also considering directing the Navy to start all over again without being limited to derivatives of either the F-16 or F-17 in order to achieve the original goal of a low-cost air combat fighter.

The Chief of Naval Operations felt compelled to write a letter to the Chairmen of House Armed Services Committee and the Senate Appropriations Committee defending the F-18 program. Dated 8 July 1975, it concluded with:

In summary, acquisition of the F-18 is in the best interests of the Navy and I believe, in the best interests of the United States. It is affordable. In conjunction with the F-14, it meets our operational requirements at lowest cost. I solicit your support of this very important Navy program.

Notwithstanding LTV's best efforts and support that their protest received from some in the Navy, in Congress, and industry (particularly Grumman), on 1 October 1975 the GAO issued its report which concluded that "the Navy's actions were not illegal or improper and that therefore the protest must be denied." It did note however, that:

The Congress has manifested significant interest in DoD's LWF/ACF programs and has closely monitored the Navy's attempts to develop a lightweight, low cost fighter that could operate effectively from aircraft carriers. The statement in the conference report on the 1975 DoD Appropriation act that "future funding is to be contingent upon the capability of the Navy to produce a derivative of the selected air force air combat fighter design" suggests that the Congress will be closely scrutinizing the Navy's choice before full-scale development funds will be provided. Thus, the ultimate determination regarding further F-18 development has yet to be made.

Fortunately for McDonnell, Congressional leadership had lost interest in forcing the Navy to buy the Vought airplane, which, given the changes that the Navy had required, was no more of an F-16 than the F-18 was. Nevertheless, industry, the F-14 supporters, and interested bystanders weren't done with trying to kill the F and A-18 program. Later that month, George Spangenberg—recently retired as director of the NavAir's Evaluation Division after 40 years as a Navy civil servant—savaged the program in testimony to the Senate Appropriations Committee:

Summing up the fighter case, the F-18 has no more capability than an F-4 and costs more; while it has far less capability than an F-14 which costs no more, and is available years earlier. There is no way in which the F-18 can be justified as a Navy fighter.

[As an attack airplane, the F-18's] true range characteristics can better be gauged by noting that it is inferior on internal fuel and without combat power usage to what was initially estimated for the A-4 in 1952. It will be recalled that the A-7 was justified in part by the fact that its capability was twice that of the A-4. Although there are other deficiencies in the design as it has been reported, its range performance alone is sufficient to disqualify it for serious consideration as an A-7 replacement. With a 50 percent higher price and a 50 percent lower capability than the A-7, the F-18 cannot be justified as an attack airplane.

Summarizing, it is clear that the F-18 is neither effective, nor cost effective, in either fighter or attack roles. It is vastly inferior in capability to the F-14 at about the same total cost, somewhat less capable and considerably more expensive than the F-4 and is inadequate in range and more costly than the A-7. The F-18 would have failed to survive any of the cost effectiveness studies conducted by the Navy in seeking F-4 and A-4 replacements in the last 15 years. There is no justification for continuing the program.

Recognizing that the A-7E did not have the maneuverability and acceleration of the A-18, Vought proposed a twin-engine, non-afterburning derivative of the A-7 that used the F/A-18's basic engine. (Vought Aircraft Heritage Center)

As damning an opinion as it was, particularly from a knowledgeable and credible source, his testimony and that of others did not even wound, much less kill, the program.

Vought's protest had been denied, but its management didn't give up trying to get the F and A-18 program canceled. In May 1977, Vought proposed an A-7E modernization, its Model 529D, powered by two non-afterburning General Electric F404 engines. The engine change would correct the major A-7 deficiency, which was thrust to weight, while not significantly degrading payload and range. It was also intended to permit the Speaker of the House, Tip O'Neill, and Senators Ted Kennedy and Ed Brooke, all Massachusetts politicians, to stay on the sidelines since both the F-18 and A-7 would be powered by GE engines built in Lynn, Massachusetts. Vought added Grumman to the team, ostensibly because of its twin-engine experience, but more to insure their support within the Navy. McDonnell was to be mollified by a contract for the AV-8B for the Marine Corps. Significant cost savings were projected.

Retired and active-duty Navy officers and civilians periodically weighed in on the side of the A-7 or against the A-18. Even inconsistent support within the highest levels of the Navy had no discernible effect. According to George Spangenberg: "By 1977 the Navy had changed direction again and Under Secretary Woolsey finally recommended to Secretary of Defense, then Harold Brown, that the F-18 program be canceled. The Navy would go with the F-14, A-7 option and the Marines would go with F-14 and the AV-8 Harrier.[8]"

Although much was made of the A-18's perceived lack of endurance, payload, and range relative to the A-7, Hornet supporters

in the Navy worked hard to defend it. First it had the advantage of more thrust and not just in afterburner, a recognized A-7 shortcoming. It had two engines, although engine failure doesn't seem to have ever been a significant worry in the A-7 community. And directly addressing the range issue, F-18 proponents claimed that flown to an optimum profile (higher and faster than the A-7), it had a radius of action as good as the A-7 with the same bomb load.[9] One aspect of this was that the A-18 needed less thrust, and therefore less fuel, when doing similar maneuvers in the target area.

The Navy initially planned to buy slightly different configurations of the Hornet, one as a dedicated fighter, F-18, and the other for attack missions, A-18. However, when the differences became fewer during the requirements finalization process, the two configurations merged into one, to be called the F/A-18 at the request of the Deputy Chief of Naval Operations (Air) in September 1978.[10] Hornet squadrons would also receive a new designation, VFA, to reflect a dual mission assignment, fighter and attack. There were concerns that the average Naval aviator could not be proficient at multiple and dissimilar missions, particularly in a single-seat airplane. Many fighter pilots, some of whose predecessors had bemoaned the loss of fuel that an onboard systems operator represented, now viewed with misgivings the loss of the extra brain and pair of eyes that the second crewman had provided in the F-4.

The first of 11 preproduction F/A-18s flew in November 1978. Flight test development was relatively untroubled, but the planform required changes, unexpectedly since it was based on the YF-17 prototype. The McDonnell-instituted snags had to be eliminated from both the wing and horizontal tail leading edges, for example. The most significant surprise, and one which stemmed from the redesign of the structure, was a roll-rate problem due to wing torsional bending

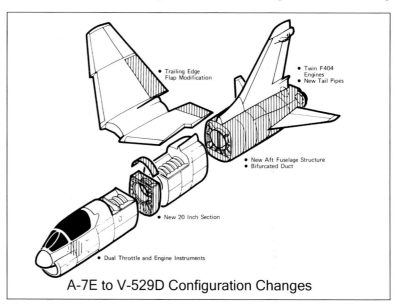

A-7E to V-529D Configuration Changes

The twin-engine version of the A-7E was projected to be relatively inexpensive, particularly if an existing airframe was used to create one. (Vought Aircraft Heritage Center)

reducing the effectiveness of the ailerons. This was solved by stiffening the wing torque box, increasing the size of the ailerons, and adding leading- and trailing-edge flap deflection when a roll was commanded. The malleability of computer-generated control surface movement was also taken advantage of to decrease the nose wheel liftoff speed by toeing in the rudders when the landing gear was down. Since the vertical fins were angled outward, this deflection provided the last increment of nose-up pitching moment needed. The digital flight-control software was also modified to provide more control authority for spin recovery after an F-18 was lost in Navy test at Patuxent River in November 1980 during an initial operational test and evaluation flight for familiarization with air combat maneuvering.

The first production Hornet flew in April 1980 as unit price problems began to emerge. As a result, in 1981 Vought again proposed two different A-7 upgrades to the Navy as the A-7X, the "Next Generation Corsair." One was powered by two non-afterburning GE F404 engines, basically the same installation as the A-7E upgrade proposed in 1977. The other was a single-engine airplane, powered by an afterburning GE F101 Derivative Fighter Engine, which looked very much like the F-8 Crusader that the A-7 was derived from. Empty weight was increased by about 1,000 pounds, but maximum catapult weight was increased by 4,000 pounds, for a 3,000-pound increase in useful load. Both were to have a modest upgrade to the A-7E avionics suite (including MFDs from the F-18) and a nose-mounted FLIR replacing the pod. The A-7X (F404) had about the same payload/range as the A-7E and increased performance and agility. The A-7X (F101) mission radius with the same payload was 20 percent better than the A-7E's and had a top speed of Mach 1.4 in afterburner, addressing the acceleration and top-speed performance advantage of the F/A-18. (And, not incidentally, it used the engine that the Navy wanted to buy to replace the troublesome TF30 in the F-14.)

In part because of Marine Corps support, and the successful passing that year of the operational evaluation of the Hornet as a fighter to replace the McDonnell F-4, the F/A-18 program survived this challenge as well.[11] The number of Hornet proponents also increased after the first F/A-18s were delivered to VFA-125 fleet training squadron in 1981—compared to the A-7, it was a hot rod. Pilots agreed that the cockpit displays and controls provided enough workload reduction that single-seat and multiple-mission operation was acceptable.

A FLIR pod, with laser designation and range finding, was developed concurrently with the basic airplane. It was first flown on a T-39 in November 1980 and made its first flight on a Hornet in August 1981. Production deliveries began in December 1983. The 340-pound

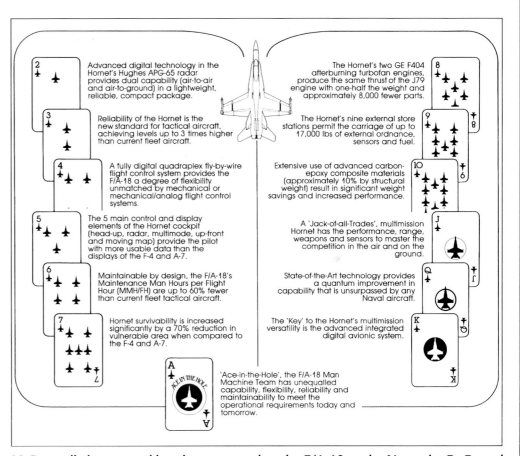

McDonnell also created brochures to market the F/A-18 to the Navy, the DoD, and Congress with emphasis on its strengths relative to the F-4 and A-7 that conveniently turned out to number 13 in this instance. (Jay Miller collection)

First flight of the first preproduction aircraft was accomplished in November 1978. The large, double-slotted trailing edge flaps, drooped ailerons, and leading-edge flaps provided a reasonable approach speed for the relatively high wing loading of the F/A-18. (Author's collection)

The F/A-18 cockpit featured three multifunction displays, a heads-up display not visible here, and hands-on-stick-and-throttles controls. These provided a fusion of information and action that allowed for one pilot to manage the workload of both fighter and attack missions. It helped that most F/A-18 pilots had video game experience. (Jay Miller collection)

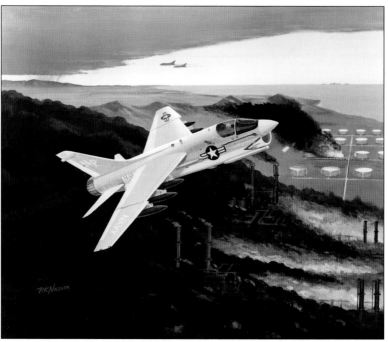

The final Vought attempt to derail the F/A-18 program was the proposal of the A-7X, which bore a striking resemblance to the F-8 Crusader. This alternative was powered by a single afterburning General Electric F101 engine. The FLIR pod hardware has been relocated to a lower forward fuselage installation, freeing up a pylon. (Vought Aircraft Heritage Foundation)

The initial at-sea carrier qualification was accomplished aboard America in November 1979. A second period of more extensive at-sea trials was accomplished aboard Carl Vinson in April 1982. One of those launches is shown here. The unique oval-shaped fuel tank was initially required for clearance of the catapult shuttle when one was installed on the centerline station. (U.S. Navy via Terry Panopalis)

pod was half the size of the A-7's and mounted on the left Sparrow fuselage station. The FLIR function provided a thermal image to one of the multi-function displays. A pod containing a laser tracker/strike camera or a thermal imaging navigation set (TINS) pod could be carried on the right cheek station. These functions were subsequently added to a new left-side pod.

In an attempt to quiet the payload/range critics in 1981, a simulated strike mission was flown unrefueled from Pax to the target range at Pinecastle AFB near Orlando, Florida, and return. The Hornet carried four 1,000-pound bombs, three 315-gallon external tanks, two Sidewinders, a FLIR pod, a laser spot tracker/strike camera pod, and 570 rounds of ammunition. The official press announcement noted that it was a full-scale development airplane with 700 pounds of fuel less than the production airplane, but it had still flown 1,240 miles with 4,000 pounds of bombs delivered at the half-way point, with enough fuel for 10 minutes of loiter upon return and a touch-and-go landing before the final landing. On 8 December, the Defense Systems Acquisition Review Council recommended that the F/A-18 be approved for full production as an attack airplane. (It had already made the recommendation for the fighter mission in June.)

The operational evaluation of the F/A-18A as an attack airplane didn't go as well. Completed in 1982 by VX-4 and VX-5, it faulted the Hornet as missing the range specification; falling short on cycle time[12]; having an inadequate fuel reserve when returning to the carrier with ordnance due to maximum landing weight; and requiring more WOD for launch than possible with some of the older, non-nuclear carriers. With respect to these shortcomings, McDonnell countered that the operational evaluation was flown at non-optimum speeds and altitudes; the landing weight had subsequently been increased; and that the WOD requirement was being reduced by airplane and catapult changes. It also claimed "the A-7 was designed for long range with a heavy war load, while the F-18 is intended for precision delivery of a smaller bomb-load, but greatly increased speed and survivability."[13]

The OpEval was not all negative. The F/A-18 did well in "bombing accuracy, radar performance and reliability (100 sorties without failure or maintenance), air combat capability and multi-role flexibility, overall reliability and maintainability, engine performance, and aircraft availability."[14]

Supporters on both sides continued to make claims of superiority using specific weapons and fuel loads. For example, the F-18's combat radius on internal fuel (which was slightly more than the A-7's) with two 1,000-pound bombs was cited as being only 220 nm versus 400 for the A-7. But F-18 proponents pointed out that with two 315-gallon external tanks and taking into account its stores pylon arrangement, the Hornet could carry six 1,000-pound bombs, two Sidewinders, and two Sparrows the same distance as the A-7 could carry six 1,000-pound bombs, two Sidewinders, and no Sparrows, which the A-7 couldn't employ anyway. Plus, the Hornet had much better acceleration and speed for maneuvering at the target and for self-defense.

In any event, the decision to replace the A-7 with the F/A-18 stood, if for no other reason than the need to replace the F-4 with an aircraft less expensive than the F-14 to operate. An F-4 replacement in the Navy

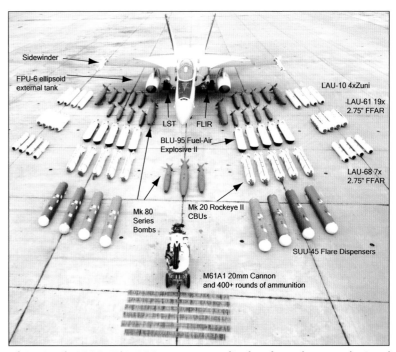

This April 1980 F/A-18A weapons display has the usual visual impact but includes some odd types and quantities of stores. For example, even with no external tanks, the Hornet could only carry a maximum of five of the eight SUU-45 flare dispensers shown in the front row and would likely carry only one on a mission, if it ever did. Likewise, it could only be loaded with half of the BLU-95s shown, although in any event that weapon would be canceled the following year. For some reason no precision weapons are included although the LST and FLIR pods are just barely visible on the fuselage Sparrow stations. (Jay Miller collection)

For carrier trials on Eisenhower, F/A-18A #3 now has the wing and horizontal tail changes resulting from flight-test problems with roll rate at high speeds and pitch-control power at low speeds. The "snag" on the wing has been trimmed off, and the one on the stabilator is filled in. (Jay Miller collection)

The AN/AAS-38B targeting FLIR shown here was carried on the legacy Hornet's left cheek station. It included the laser spot tracker (LST) feature of the AN/ASQ-173 laser detector tracker/strike camera pod, which was carried on the right cheek station, in addition to a narrow field of view (FOV) FLIR and laser target designator/rangefinder. The -38B pod was replaced by a substantially improved AN/ASQ-218(V)1 advanced targeting FLIR (ATFLIR), which also included a wide FOV navigation FLIR on its fairing that replaced the alternate pod previously carried on the right cheek station, the AN/AAR-50 Thermal Imaging Navigation Set (TINS). When a Targeting FLIR was carried, the aircraft fuel tanks were normally mounted on the centerline and right wing to maximize the pod's field of regard. (Kevin D. Austin)

The primary objective of the study was probably to pressure McDonnell to cut its price, since the investment to set up a second source would almost certainly more than offset any price reductions gained from competition, not to mention the loss of learning curve benefit that would result from splitting the total quantity between two production lines. In any event, no second source resulted.

The Hornet fell short in more than range compared to the original contract: [15]

	Speci-fication	FSD	Lot 14
Fighter Radius (nm)	420	319	302
Strike Radius (nm)	618	437	398
Tanks (#/gal)	3/300	3/330	3/330
Specific Excess Power (ft/sec)	753	617	584
Acceleration from Mach 0.8 to 1.6 (seconds)	98	144	180
Minimum Carrier Approach Speed (kt)	128	140	142

What's more, its endurance compared to other aircraft required careful attention during cyclic carrier operations. (Because of the limited deck space, an aircraft carrier generally operates to a launch/pull forward/land/pull aft cycle, which means it is only available for landings every 90 minutes or so except in an emergency.) The F/A-18 was derisively referred to by some as the lawn dart, because when you throw one up, it comes right back down. As a result, in 1985 Lehman requested a budget for Navy land-based tankers to support carrier operations in the Mediterranean. The Senate Armed Services Committee refused to authorize a tanker force separate from the Strategic Air Commands. Lehman's concern was realized in late 1985, when *Coral Sea* F/A-18s on a strike exercise almost ran out of fuel because Air Force tanker support was allegedly withdrawn mid-mission. The flight was saved by an emergency launch of KA-6Ds from *Saratoga*.

While not caused by the F/A-18's lack of range, per se, the *Coral Sea* incident was used to revisit the Hornet's range/endurance shortcoming. According to the *Armed Forces Journal* article describing it, "During the *Constellation's* first deployment with F/A-18s…somewhere between 70 percent and 93 percent of all A-6 sorties were flown for refueling missions. Normally, only about one-third of all A-6 sorties are refueling missions." The same article reiterated the comparison that the F/A-18 had only half the mission radius of the A-7 when loaded with four Mk 83 bombs.[16] However, some of the criticism was voiced by individuals in the medium attack community and others in the Navy who saw the proliferation of the F-18 as threatening the deep strike mission. They were abetted by Grumman, which was trying to protect and extend its F-14 and A-6 production.

None of the many attempts to derail the F/A-18 program were successful. It benefited from the increasing use of guided and powered weapons more than an A-7 would have. Only a couple of them needed to be carried to deal with a target, so the need for external tanks was

was also necessary as long as the Midway-class carriers were still being deployed, since the jet blast deflectors had not been upgraded or the avionics support added to accommodate the F-14, among other things. *Franklin D. Roosevelt's* last deployment ended in April 1977, but *Coral Sea* was to keep deploying until September 1989 and *Midway,* August 1991. With the F/A-18 available, the last F-4 deployment aboard *Coral Sea* ended September 1983 and on *Midway,* May 1986. However, since *Coral Sea* was in overhaul, the first F/A-18A deployment began in February 1985 aboard *Constellation,* with two squadrons replacing the A-7s. *Coral Sea* deployed in October 1985 to the Mediterranean with four F/A-18A squadrons (two being from the Marine Corps) and one A-6 squadron, arriving in time for a confrontation with Libya.

John Lehman had become Secretary of the Navy in 1981 in the Reagan administration. He became one of the most involved and hands-on SecNavs in history. He was also one of the most knowledgeable from a Naval aviation standpoint, since he was an A-6 bombardier/navigator in the reserves. In 1985, Lehman, facing a budget crunch, requested a study of alternatives for competitive procurement of the F/A-18, which would conceivably result in a price reduction. Lehman had surely noticed that McDonnell's profits for 1984 had risen 28 percent on a sales increase of 11 percent. Northrop was one of the logical candidates, since they already produced the center and aft sections of the F-18 fuselage and were a prime contractor as well.

Flow visualization, as provided by a Blue Angels F-18 at high speed and angle of attack in humid conditions. Note the tight vortices coming off the inboard and outboard ends of the leading edge slats and the more diffused one coming off the leading edge extension (LEX). In 1984, problems with vertical fin attachment fitting cracking were found. The higher-than-predicted loads were caused by the LEX vortex, which is intended to delay wing stall at high angles of attack. A beef-up was installed at the base of the vertical fin. A small fence was also added on the LEX to adjust the vortex characteristics to reduce the loads on the tail and still provide the lift benefit. (Author's collection)

The Coral Sea returned to the Mediterranean in October 1985 on its first deployment with F/A-18 squadrons just in time for a confrontation with Libya. The Hornets were used to provide combat air patrol during Operation Prairie Fire and attack Libyan air defense radars and missile sites with high-speed anti-radiation missiles during Operation El Dorado Canyon on 15 April. (Jay Miller collection)

In order to more than double the AGM-84E SLAM's standoff range, wings were added to create the AGM-84H SLAM ER (for expanded response) shown here in the NOTS China Lake Museum. It had wings for better range, a more potent warhead, and software changes to improve designation of the aim point. The wings were hinged so they could be stowed under the missile while it was being carried on a wing pylon. The first SLAM ER launch from an F/A-18C was in March 1997. SLAMs in inventory could be retrofitted to the ER configuration. (Terry Panopalis collection)

not a disadvantage, and they had more range, so the F/A-18's was not such a handicap. It also had its good points. It was reliable and the maintenance man-hours per flight hour were low compared to the F-14, F-4, A-6, and even the A-7. Its initial accident rate was remarkably low: after a total of 154,000 hours in service, it was 5.2 per 100,000 hours. By contrast, the A4D had an initial accident rate of almost 60 and the A-7 had 32.9. Fighter pilots were able to achieve acceptable levels of weapons delivery accuracy after minimal training and attack pilots appreciated the acceleration and maneuverability that afterburning engines provided. Much was made of an engagement in 1991 on the first day of Desert Storm when two pilots flying F/A-18s, each loaded with four 2,000-pound bombs, shot down two Iraqi MiGs and continued on to bomb the target.

McDonnell produced 370 production F/A-18As and 40 two-seat F/A-18Bs (originally designated TF/A-18A) before production transitioned in 1987 to the single-seat F/A-18C as well as the two-seat F/A-18D in 1988 for the Marine Corps. The 465 Cs incorporated avionics upgrades to increase night and all-weather attack capability and employ the AIM-120 Advanced Medium Range Air-to-Air Missile (AMRAAM), the AGM-65F Maverick air-to-surface missile

On 26 October 2001, as part of an early morning bombing mission in support of Operation Enduring Freedom, a VF-192 F-18C Hornet from Kitty Hawk has just finished refueling from a USAF KC-135R. (U.S. Navy 011026-F-4884R-006)

An unexpended AGM-88 High-speed Anti-Radiation Missile (HARM) on this Hornet has just been dearmed immediately following landing by Aviation Ordnanceman 3rd Class Quentin Bryant, who is so indicating as he clears away. "Ordies" wear red to distinguish them from other specialists on the flight deck. (U.S. Navy 980219-N-0507F-001)

F/A-18C/D
FY86 Block Upgrade

Adds:

- I²R Maverick (AGM-65F)
- AIM-120 AMRAAM Provisions
- ALQ-165 (ASPJ) Provisions
- Improved Maintenance Monitoring (FIRAMS)
- RECCE Common Nose Provisions
- Independent L/R Fuel System
- LTD/R Provisions
- Data Storage Set

The F/A-18C was a block upgrade of the basic F/A-18A. Some of the improvements were retrofittable to the F/A-18A, which was then known as an F/A-18A+. The D was the two-seat version of the C. It was primarily operated by the U.S. Marine Corps and further upgraded for improved night attack capability. (Jay Miller collection)

with infrared seeker, and the AGM-84 Harpoon anti-ship missile. The last of the F/A-18Cs was delivered in 1999, having replaced many of the F/A-18As and the remaining A-7Es in deployable squadrons beginning in 1988. During and following production, both the As and the Cs were frequently upgraded for increased mission capability. For example, many As were retrofitted with the

APG-73 radar and other F/A-18C features; these were designated F/A-18A+ and primarily assigned to Marine Corps and the two Navy reserve squadrons.

As it turned out, trials and tribulations of the F/A-18 program were almost inconsequential compared to what was to come with the attempts at A-6 replacement.

Although the new generation of attack airplanes had less range than its predecessor, the shortfall was offset by the new generation of missiles like the AGM-84E Standoff Land Attack Missile (SLAM) shown here, which was basically a stretched Harpoon with a different targeting system. (Jay Miller collection)

The pilot of a VFA-136 F/A-18C tests its flare countermeasures system prior to heading into Iraq in August 2007. In addition to the 20mm cannon, the Hornet is armed with one Maverick missile, one JDAM, and one laser-guided bomb to provide a tailored response to close air support requests. (U.S. Navy 070817-N-6346S-380)

CHAPTER ELEVEN

REPLACING THE A-6, ROUND ONE

This TBM-3D was photographed in April 1945 while configured with a set of sealed-beam lights around the cowling and on the wing leading edges. It was an example of experiments to reduce observability named Project Yehudi after a popular radio show's running joke at the time about "the little man who wasn't there." The concept was to reduce the contrast of the aircraft with the sky behind it during daytime operations so it couldn't be seen by the target it was approaching until it was at close range. After fine tuning of the light intensity, it worked as intended, with the detection range reduced from about 12 miles to less than 2, but wasn't considered effective enough for service use. (National Archives 80-G-411528)

In the early 1980s, Navy planners were wrestling with how to upgrade or replace both the F-14 and the A-6. Both were expensive to maintain. The F-14 had been saddled with an unsatisfactory engine for a decade. The A-6 was aging both structurally and from a mission avionics standpoint. As early as 1981, the Navy projected that the A-6s would have to begin to be replaced by the mid-1990s. The solution was to be defined by the study of a VFMX, an advanced multi-mission, carrier-based fighter. However, in March 1986 the Navy and the Air Force eventually agreed to an OSD-facilitated and Congressionally imposed plan that the Navy and Air Force would develop the Advanced Tactical Aircraft (ATA) and the Advanced Tactical Fighter (ATF), respectively.[1] The ATA would replace the A-6 and the F-111 while the ATF would do the same for the F-14 and F-15. The Navy would bridge to the new aircraft with A-6 and F-14 upgrades. From the Navy's standpoint, the result was the expenditure of billions of dollars with nothing to show for it but some mission avionics and engine and airframe technology.

The organizers of an air show at NAS Whidbey Island, the home of the West Coast A-6 squadrons, arranged this A-12 static display. Little did they know how accurate an observation this was on the status of the program and that this was all that they would ever see of the A-12. (Darryl Shaw via Dennis Jenkins)

Stealth

Radar-assisted defenses had become a real problem by the late 1960s. Extremely low-altitude ingress provided a measure of survivability but severely penalized the acquisition and attack of some targets and risked running a gauntlet of massed antiaircraft emplacements and man-portable air-defense missiles, e.g., the Russian SA-7 Strela, that homed in on the engine exhaust plume. Jamming, chaff, and flares provided some protection on ingress and egress, but the presence of jamming was an indication that a raid was imminent. In close proximity the enemy radar might be able to distinguish the attacking aircraft from the clutter, sooner if the defense included radar capable of frequency hopping. And the infrared-guided missiles were increasingly capable of distinguishing between a flare and engine exhaust.

Since radar was the primary means of aircraft detection and targeting, one solution was a cloaking device that would absorb the radar emission and/or deflect it in any direction except back at the transmitter. Such a capability was becoming increasingly desirable. The U.S. had first-hand experience with Soviet SAM technology in the Vietnam War. The Israeli Air Force had a very difficult time in the 1973 Yom Kippur War, losing more than a hundred airplanes in less than three weeks to an improved SAM and anti-aircraft fire.

In 1974, the Defense Advanced Research Agency (DARPA) initiated a program to study low observable technology as it applied to aircraft, with the objective of reducing radar, infrared, noise, electronic, and visual signatures as much as possible.[2] A low radar cross section (RCS) was the highest priority. After initial studies and an RCS "pole-off" of Northrop and Lockheed models, the Lockheed Skunk Works received a contract in April 1976 for Have Blue, the flight test of two J85-powered XSTs (Experimental Survivable Testbed). The first, the flying qualities demonstrator, flew on 1 December 1977. It physically

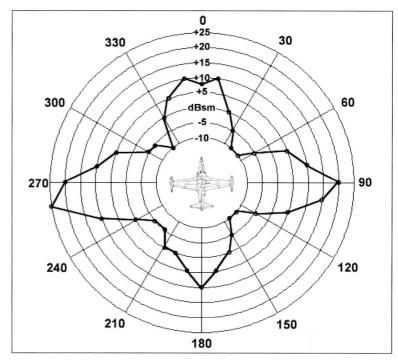

This is an example of the significant variation in radar cross section at 10-degree increments around the azimuth of an airplane, in this case a T-33 trainer, (seen here in Navy markings, which denotes the jet as a TV-2), for a specific radar frequency. The fuselage sides reflect the most energy. The peak from the aft end is primarily the engine's turbine and the two slight peaks from in front are the engine inlets. Similar variations can be expected at different viewing angles, such as looking slightly up at the aircraft as a ground-based radar would. (Author's collection)

represented the aerodynamically challenged low-RCS configuration comprised of flat surface areas and a highly swept wing calculated to deflect, not reflect, the radar pulse.[3] The second aircraft was covered with radar-absorbing material to further decrease the radar cross section. It flew for the first time in July 1978.

The reduction in radar detectability was fantastic. Radar cross section (RCS) can be defined as the equivalent of a metal sphere. If the RCS of a strike fighter without stealth shape/treatment is the same as a sphere

In order to minimize the radar return, the F-117 was faceted with the angles carefully chosen to not reflect energy back at the radar. The engine compressors were masked with a grillwork and the exhausts of the non-afterburning engines were shielded from direct observation except from a narrow viewing angle above the aircraft. In addition to shaping, the aircraft is built and coated with radar-absorbing materials. (Jay Miller collection)

of 5 to 10 feet in diameter, the RCS of a stealthy one would be less than an inch in diameter—from some directions at some frequencies.

The demonstration program was successful from both handling qualities and radar cross section standpoints. Detection proved difficult except at ranges far too short to provide an effective defense, with lock-on of fire-control radar unlikely. The Senior Trend program resulted. A contract was awarded to Lockheed in November 1978 for five prototypes of a small bomber designated the F-117A. Since existing aircraft radars would have compromised stealth, target acquisition and designation was accomplished with a pair of infrared imagers/designators. This kept the aircraft from being detected by its use of radar, but resulted in a visibility requirement for employment of its only weapon at the time, the laser-guided bomb.

The first F-117 flew on 18 June 1981. Low-rate production of operational aircraft was accomplished in parallel with the develop-

ment program. The first of 59 production F-117As was delivered in 1982 with limited operational capability achieved in October 1983. The existence of the F-117A was finally publicly revealed on 8 November 1988. Its first announced operational use was in Operation Just Cause, the invasion of Panama in December 1989 to arrest its leader, Manuel Noriega. Eight F-117As flew a non-stop 6,000-mile round trip with air refueling from the base in Tonopah, Nevada, to bomb air defenses in the Canal Zone.

From a practical standpoint, the minimal RCS could not be maintained from all aspects. Further, as the range decreases, the detectability increases because the strength of the return increases. The surface of the stealthy aircraft was also high-maintenance, as even a minor "flaw" such as a scratch or fastener head protruding, not to mention an ill-fitting door, could cause the signature to blossom, relatively speaking. The F-117A was therefore not undetectable on radar, but

Aviation Ordnanceman 3rd Class Allen Crow prepares to move a 2,000-pound GBU-32 Joint Direct Attack Munition (JDAM) for transport to the John F. Kennedy's flight deck. The John F. Kennedy and her embarked Carrier Air Wing Seven (CVW-7) were conducting combat missions in support of Operation Enduring Freedom in 2002. (U.S. Navy 020603-N-6913J-001)

The B-2 was a second-generation stealth design. Although facets are still important, as seen by the repetition of leading and trailing edge angles, improvements in surface reflectivity analysis and materials allow a more rounded and aerodynamic shape to be used. (National Archives)

close enough to it to be nearly invulnerable under certain conditions. When it was approaching, for example, search radars could catch fleeting glimpses, but no tracking radar could hold it for anywhere near long enough to shoot one down.

Unmarked in more than 1,300 missions during the 1991 Gulf War, a SAM finally shot down an F-117A in Serbia in March 1999. The pilot ejected and was rescued. The success of the Serbs was a result of a combination of carelessness on the part of the USAF mission planning (the same route was reportedly used four nights in a row) and creative use of all-but-obsolete Soviet tracking radar and missiles by a Serbian battery commander. However, subsequent strikes in 2003 that began the invasion of Iraq were again accomplished without the loss of an aircraft or apparently even any damage. The remaining airplanes were retired in 2008.

Northrop B-2 Program and JDAM

The Northrop B-2 program was initiated in 1981. A long-range stealth bomber, it was both much larger and more technically ambitious than the Lockheed F-117. Among other advancements, its shape was stealthy without having to be faceted like the F-117's and it incorporated Low Probability of Intercept (LPI) radar that included synthetic aperture for precision targeting capability of the new Joint Direct Attack Munition (JDAM). It was publicly revealed in November 1988 even before its first flight, which did not finally occur until July 1989. The first operational aircraft was delivered in December 1993, but it wasn't fully mission capable due to delays in development

Grumman's marketing department went all out on the paint scheme of the first A-6Fs. Although this first one has a metal Grumman wing instead of the Boeing composite wing and does not have the mission avionics, it does have the extra pair of wing stores stations for air-to-air missiles and the additional cooling scoop on the aft fuselage in front of the vertical fin so it was aerodynamically representative of the F. (Grumman History Center)

of the defensive avionics system. However, it was combat ready for the conflict in Kosovo in 1999, also dropping the JDAM for the first time in combat. The B-2s involved flew a 30-hour, non-stop round trip from Whiteman Air Force Base in Missouri, their range extended by inflight refueling.

JDAM eliminated the need to visually acquire a target as required by laser-designated bombs. A JDAM was created from an ordinary Mk 80 series bomb by substituting a guidance and control tail section for the bomb's usual sheet metal afterbody. A girdle with low-aspect ratio fins, or strakes, was also wrapped around the center section to provide for increased lift and therefore better cross-range performance. The JDAM tail section incorporated a small computer, inertial measurement device, and Global Positioning System (GPS) sensor that together guided the bomb to the set of geographic coordinates provided as the target before release. Primary guidance was provided by the GPS unit, with inertial measurement providing a backup capability if the GPS signal was lost or jammed. Range depended on the release altitude or toss angle/speed, but it could be up to 15 nm with no degradation of accuracy with distance. CEP with GPS/INS guidance was about 30 to 40 feet. If position correction was provided from a ground station (Differential GPS) or the bomber's radar, accuracy was improved to several feet, almost as good as a laser-guided bomb.[4]

Moreover, JDAM did not require target designation like the laser-guided weapons and was autonomous after release. There could be clouds or other limitation of visibility between the airplane and the target. Once the bomb was dropped, the airplane was free to not only maneuver but also to bug out. JDAM was also cheaper than the laser-guided weapon. However, as with email, it would not be delivered to the intended recipient unless the address was exactly right. If the target coordinates were incorrectly determined, transmitted, or entered, it would not only miss the intended target, but also precisely hit whatever was at the coordinates it had been given, if it could get there. Errors were not unknown: in 2001 in Afghanistan, a JDAM dropped by a B-52 was inadvertently given the coordinates of the allied ground unit itself by the attack controller on the team. It could also not be used against a moving target as a laser-guided bomb could. JDAM capability was subsequently updated by adding a laser seeker to the nose of the bomb, which allowed it to be guided by laser designation when possible and required.

Although the Air Force planned to buy at least 132 B-2s, in the end only 21 were produced, including the six development test aircraft that were eventually brought up to an operational configuration.

Grumman A-6F Intruder II Program

As a result of ongoing structural problems and high maintenance cost, in the early 1980s the Navy was planning to replace the A-6E. In 1983, the Navy had two proposals in hand, one from McDonnell Douglas to upgrade the F-18 and another from Grumman for a similar effort on the A-6. In the case of the F-18, McDonnell needed to make it bigger and significantly improve its all-weather attack capability. The A-6 needed a major upgrade to its entire avionics suite.

Both were projecting that the cost would be within the Navy's $500 million R&D budget.

The McDonnell proposal for a two-seat A-18 (AW) increased the F-18's wing area by 32 square feet and incorporated a flap modification to accommodate a 13 percent increase in gross weight to 55,270 pounds, including the capacity for 3,000 pounds of additional internal fuel and two 460-gallon external tanks. The F404 engine thrust would also be increased by 15 percent. Automatic terrain-following radar and offset-bombing capability were to be added along with larger rear cockpit displays. The result was a high-performance airplane compared to the A-6 with a self-defense capability—air-to-air radar and Sparrow/Sidewinder missiles.

Grumman's upgrade of the A-6 resulted in an empty weight increase of only about five percent. The higher weight would be more than accommodated by the use of the EA-6B wing fillet and longer slat and either of two engine options, the 11,200-pound thrust Pratt & Whitney J52-P-408 that powered the EA-6B or a non-afterburning General Electric F404 with slightly less thrust but still a 15 percent increase. Various radar improvements were proposed, all of which retained the A-6's simultaneous terrain clearance and tracking capability. The remainders of the avionics changes were displays and a computer common to planned improvements to the F-14D, F/A-18, and AV-8B. An extra pair of outboard pylons would be added along with Sidewinder capability. An auxiliary power unit (APU) provided a self-start capability.

The A-6F had better payload-range; the A-18 (AW) was judged to have better survivability in part because of its supersonic performance. The Navy, including the SecNav who was an A-6 B/N in the reserves, preferred the A-6F; the Marine Corps favored the A-18 (AW).

In 1984, Grumman and the supporters of the A-6 were finally successful in getting a go-ahead for the A-6F along with one for the F-14D, with some of the avionics to be common. Initially referred to as the A-6E Upgrade, it was virtually an all-new design, retaining only the basic shape and fuselage of the A-6. The engines were to be General Electric F404-GE-400D non-afterburning, low bypass turbofans, almost identical to the basic engine in the F/A-18. It would have Norden radar with both synthetic and inverse synthetic aperture modes, a new digital cockpit with five multi-function displays, and a heads-up display. The Boeing composite wing would be incorporated with a stores pylon on each outboard wing panel and modifications to increase lift including modification of the fillet/slat to extend the slat seven inches inboard. An expanded self-defense capability was provided by the new radar combined with Sidewinders or the new AIM-120 AMRAAM.

McDonnell and the Marines got a consolation prize, the two-seat F-18D, a less ambitious version of the A-18 (AW). The aft cockpit was for a weapons-system officer and was not equipped with flight controls. The 147 Ds were primarily delivered to the Marine Corps to permit the release of its A-6Es back to the Navy. Some were modified to provide a reconnaissance capability in lieu of the M61A1 cannon as a replacement for the Corps' aging RF-4Bs. Some replaced the OA-4 Skyhawks and OV-10s used for forward air control and similar

The first 31 F/A-18Ds were not fully night capable. The first F/A-18D was eventually used as the prototype for the Night Attack F/A-18D, flying in this configuration in May 1988. (Jay Miller collection)

missions. The aircraft was carrier-qualified so Marine squadrons operating F/A-18Ds could deploy from aircraft carriers.

One of the major deficiencies of the A-6 that was difficult to address was its lack of stealthiness. Grumman had made an attempt to apply radar cross-section reduction technology to the A-6 but Secretary of the Navy John Lehman, at least, was unimpressed. "I thought Grumman was way out of touch," said Lehman. "What they were attempting to do was like taking a dump truck and trying to make it pretty by adding fins." Always candid, he said as much to George Skurla, then head of Grumman's aerospace unit in July 1984.[5] Mr. Skurla subsequently visited Lockheed and Northrop and discovered to his chagrin that Lehman was right.[6]

Five A-6Fs were ordered in July 1984. Three of the five flew in various configurations, the first in August 1987 with the F404 engines but without the new avionics. The second flew in November, but that same month Congress passed the 1988 defense authorization bill that ordered the Navy to terminate the A-6F program and continue with the ATA. (Congress was not of one mind on the subject, as evidenced by funding for the A-6F being included in the 1988 defense appropriations bill.) The Secretary of the Navy, John Lehman, had resigned in April. Without his protection, the Intruder upgrade proved to be vulnerable, particularly once the ATA program was under way.

The Navy had usually insured against the failure of a new aircraft program by having a similar one in development or upgrade as well. Once the new aircraft had been determined to be a success or failure, the backup would be terminated or only procured in limited numbers. This hedging strategy was no longer viable as the cost of developing an aircraft soared along with the degree of OSD and congressional oversight of the services' programs. As capable as the A-6F was, it was inferior to the ATA in significant respects and would not be in service for very long before the ATA was deployable according to program plans.

The A-6F was to have five identical multifunction displays (MFDs) in lieu of the dedicated displays of the Intruders that preceded it. The 25 buttons and three knobs on the bezel surrounding the picture were used to manage the display and control the functions being displayed. Any of the five MFDs could be used to show any of the data that could be displayed. The pilot was also provided a large heads-up display that replaced the gunsight. (Grumman History Center)

Work on the A-6F avionics system continued for a few months under Navy contract after the A-6F itself was canceled while the benefit of substituting a less expensive airplane, the A-6G, were evaluated. This would provide the customary hedge against schedule slip or outright failure of the ATA program. The third A-6F, the first to have the digital avionics system, flew in August 1988. The A-6Gs were to be rebuilds of A-6E airframes with new wings, the A-6F avionics, and the higher-thrust J52-P-408A engines that powered the EA-6B. However, the Navy decided to end the A-6G effort in December and rely solely on the successful development and qualification of the ATA.

Grumman now had to pin its hopes of being involved in the replacement of the A-6 on the failure of the ATA combined with a successful campaign to sell Congress, OSD, and the Navy on a strike fighter derivative of the F-14.

General Dynamics/McDonnell A-12 Avenger II Program

Compared to the A-6 that it was to replace, the Navy's Advanced Tactical Aircraft was to have a significant increase not only in stealthiness but also in payload and range, approximately 5,500 pounds and 1,000 nm radius of action. It was also to be able to return to the carrier with twice the weapons load that the A-6 could. As was now *de rigueur*, a major improvement in reliability and maintainability was also required as well compared to that of the airplane being replaced—twice the reliability and half the maintenance man hours per flight

Bureau Number 162183 was the A-6F Intruder II aerodynamic and propulsion system prototype. It first flew on 26 August 1987 with the F's F404 engines but not with the F's avionics. (Grumman History Center)

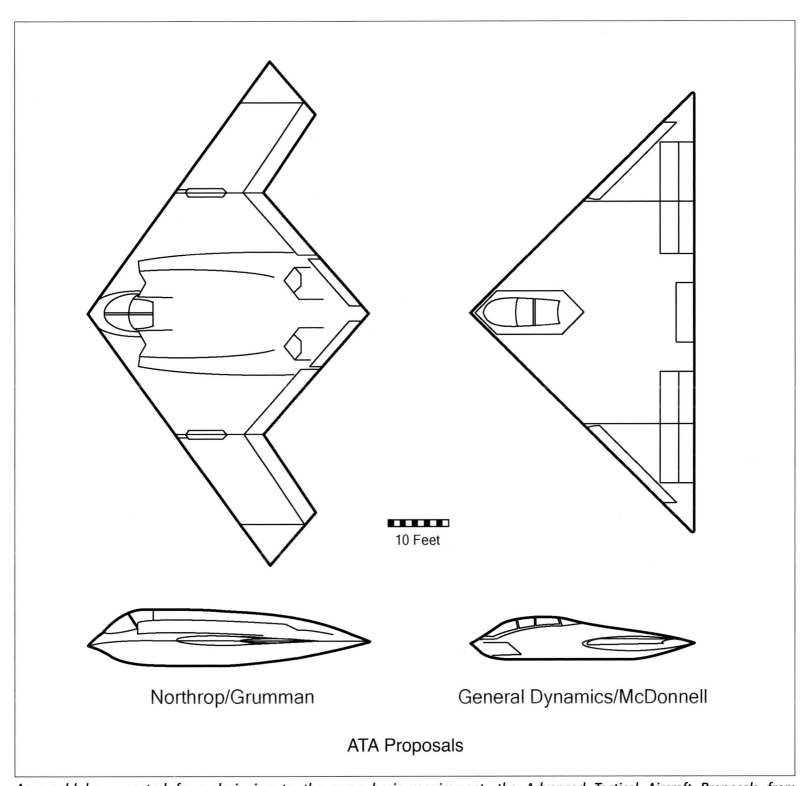

Northrop/Grumman

General Dynamics/McDonnell

ATA Proposals

As would be expected from designing to the same basic requirement, the Advanced Tactical Aircraft Proposals from Northrop/Grumman and General Dynamics/McDonnell were approximately the same size. The Northrop proposal was reminiscent of the B-2 layout with the engine intakes and exhaust on top of the wing, whereas the General Dynamics proposal was a simple triangle with the engine inlets and exhaust below it. (Author)

Because the A-12 did not have a separate horizontal tail to control the pitch change that flaps imparted, none of the surfaces on the trailing edge act as a flap. Together, they provide roll, yaw, and pitch moments as well as adding drag if required to reduce speed quickly. High lift was provided by the large wing area relative to weight and the leading edge slats, which did not cause a pitching moment when extended. Note that the edges of the end of the canopy and the edges of access panels are parallel to the wing leading edges, a stealth feature to minimize radar cross section. (Jay Miller collection)

This photo of the A-12 mockup was taken with the inlets blocked off to conceal the design features that shielded the engine compressor face. For minimum radar cross section on the first-day strike, all weapons were to be carried internally. Here, only the left-side internal weapons bay doors are open. The inboard bay was big enough to carry two Mk 84 or five stacked Mk 83 bombs. The outboard bay was for an air-to-air missile. The bay in between the two houses the landing gear when it is retracted. (U.S. Navy via author's collection)

The A-12 mockup was impressive. It included the cockpits and internal and external lighting. Green position lights on the wingtips and behind the canopy provide a visual reference for the pilots flying formation at night. (U.S. Navy via author's collection)

hour. In addition to ordnance delivery, the ATA was to carry air-to-air missiles for self defense and have a reconnaissance role. Initial Operation Capability was to be attained in 1994, which dovetailed with the Navy's A-6 and the Air Force's F-111 retirement plans.

It was to be a huge business for the winner, with the Navy intending to buy 858 airplanes, including 104 for the Marines. The Air Force was to buy 400. It was also one of the few new aerospace programs.

After about three years of preparatory effort, Northrop/Grumman/LTV and General Dynamics/McDonnell Douglas teams competed for a full-scale development contract. The aircraft were similar in basic characteristics but very different in configuration.[7]

	Northrop/Grumman/LTV	GD/McDonnell
Wingspan (ft)	80	70
Length (ft)	46	35
Takeoff Weight (lb)	69,316	69,713
Internal Fuel (lb)	24,358	21,322
Payload (lb)	5,550	5,160

Although Northrop should have been favored for its greater stealth background, all other things being equal, its team balked at the ceiling price[8] and other contract requirements that the Navy was insisting on, given the development risk involved in meeting the specification. The General Dynamics/McDonnell Douglas team was not deterred by the terms and was therefore awarded a contract for the A-12 Avenger II in January 1988. First flight of the first of eight full-scale development aircraft was scheduled for June 1990. It was not to happen.

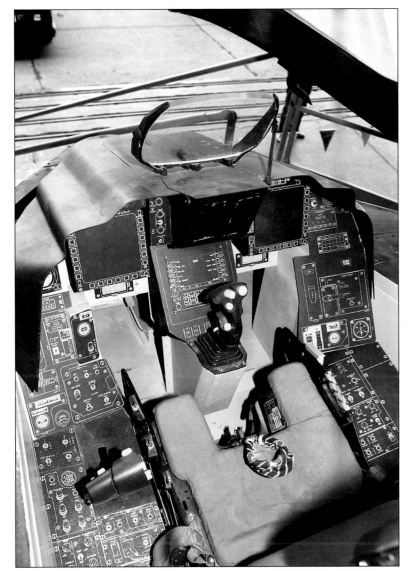

The A-12 pilot's station featured three multifunction displays, a very large head-up display, a conventional center stick but with a very small throw, and two large throttles. Note the ergonomic shape of the throttles. (Jay Miller collection)

The bombardier/navigator was provided with several displays (the one just outboard of the right-hand controller is not a portable one but the face of the center one that has come adrift) and a controller for each hand. (Jay Miller collection)

A major milestone, the Critical Design Review, was finally passed in October 1990 with first flight having slipped to early 1992. In November 1990, General Dynamics and McDonnell Douglas, realizing that they were going to overrun the development contract by at least $1 billion, and the prospects of recovering it were shrinking because of the reductions in production quantities, not to mention schedule slippage, formally requested a restructuring of the FSD and production contracts, which included fixed-price options for early aircraft. At that point, only a full-scale mockup and a few production parts existed, in addition to the avionics and components work at the many A-12 subcontractors. More than $2 billion had been spent.

The higher unit cost projected only exacerbated the problem of funding production at a reasonable rate as well as buying the other airplanes the Navy needed each year and still stay within annual procurement budget guidelines. The program was therefore deemed unaffordable, at least by OSD, regardless of its mission capability. To be fair, part of the increase was the result of OSD's decision to drop the Marine Corps procurement and delay the Air Force's, significantly reducing production rate (increasing overhead cost) and the impact of the learning curve (higher average unit cost).

Although the Navy and the contractors had reached agreement on restructuring the development and production contract from a cost

One concept for shrouding the engine compressor on the A-12 was a simple set of louvers as shown here on the mockup. A Plexiglas sheet with Plexiglas handles was placed in the inlet to protect the louvers during this public viewing at General Dynamics in Fort Worth. (Greg Fieser photograph via Dennis Jenkins)

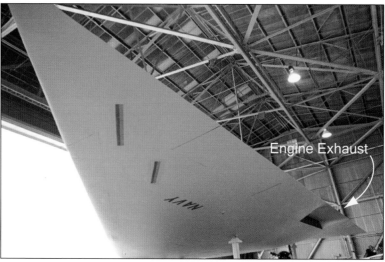

Unlike almost all other stealth designs, the A-12 engine exhaust was shielded from above, probably to minimize the IR signature presented to the air-to-threat. There were also pitching moments with power change benefits, a handling qualities concern for carrier suitability. (Greg Fieser photograph via Dennis Jenkins)

Like the NACF that resulted in the F/A-18, the NATF was to be a marinized version of the winner of the Air Force Advanced Tactical Fighter flyoff, which turned out to be the F-22. Lockheed claimed that in addition to having common engines and basic airframe as shown here, the avionics, armament, and aircraft subsystems would be 80 percent common, saving $11 billion (FY85 $) in development, production, and support costs compared to the cost of a new stand-alone Navy fighter program. (Author's collection)

standpoint, in January 1991 the Secretary of Defense, Richard Cheney (the future Vice President of the United States), ordered that the program be terminated for default "due to the inability of the contractors to design, develop, fabricate, assemble and test A-12 aircraft within the contract schedule and to deliver an aircraft that meets contract requirements."[9] There was some urgency to the decision, as the next increment of program funding was about to be obligated in accordance with the contract.

Not everyone in the Navy lamented the cancellation of the A-12 program. It was about to demand a huge proportion of the annual Navy budget. While no one would argue that low RCS was not desirable, it was becoming apparent that it was expensive to achieve at the values being attempted in the A-12 program, the effort required to maintain it at those levels was probably high, and it might not be possible to keep in a carrier-based environment in any event.[10] The A-6 program had almost floundered because of the difficulty in keeping its advanced all-weather avionics suite working. This was a similar situation.

Thirty months between contract go-ahead and first flight was acceptable for a new airplane program, given adequate preliminary design, trade studies, and wind-tunnel test. It was optimistic given the degree of difficulty of the A-12 program and the contractors' relative lack of experience with the technology involved. For the B-2 program, Northrop was given six years between contract go-ahead and first flight, compared to less than three for the A-12. Even then, Northrop missed the first flight date by more than a year, albeit partly due to a change by the Air Force to add low-altitude, terrain-following ingress.

Two of the major risks were the very high content of composite structure for weight reduction and the radar cross-section requirements. These and some of the mission avionics were not only pushing

An F-14A/B Tomcat aircraft assigned to the Strike Aircraft Test Directorate of the Naval Air Warfare Center Aircraft Division Patuxent River is shown conducting a separation test of the ADM-141 Tactical Air Launched Decoy (TALD). The testing began in November 1993 and was completed on 28 April 1994. (U.S. Navy DN-SC-95-01057)

the state of the art from both a design and manufacturing standpoint, but General Dynamics and McDonnell (with the Navy's encouragement) had all but fired them out of a cannon and were hoping to catch them down range. As a result, less than three years after contract award, the program was more than a year behind schedule and projected to overrun significantly. The estimated weight had increased as well, but that was to be expected.

It's likely that the Northrop-led team would have done a better job at execution of the program. The B-2 program had preceded the A-12 program by about six years. That experience might have resulted in fewer and less costly composite design and fabrication problems than the General Dynamics team encountered. Certainly Northrop was more realistic in its projections of the program cost and schedule, a disadvantage of being the more knowledgeable competitor.

It's unknown and unknowable whether the production A-12 would have been acceptable in performance even though it was overweight, assuming that the worst of the misestimates had been encountered and accounted for. However, there were those in the Navy who were neither shocked nor terribly disappointed about the weight and schedule status as of the end of 1990 and were aggressively promoting the program. In November 1990, VADM Richard M. Dunleavy, assistant chief of operations for air warfare, admitted that the growth in A-12 weight had eliminated it as a candidate to do Advanced Tactical System (ATS) aircraft missions (few in the VAW, VS, and VAQ communities believed that an A-12 derivative was the best choice anyway),

One interpretation of this photo is that the carrier-based Navy, represented by the A-6 Intruder, is behind in the race with the Tomahawk cruise missile, launched from the nuclear-powered attack submarine La Jolla in April 1983. The A-6 crew is in fact monitoring it on its flight to a target in the Tonapah Test Range in Nevada. As it turns out, however, the carrier-based Navy will cede much of the deep strike mission capability to Tomahawks launched from submarine and non-aviation ships. (U.S. Navy DN-SC-84-10104)

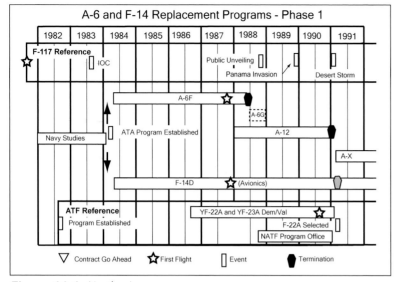

Figure 11-1. (Author)

but the A-12's current weight "is acceptable to we the operators… we can work with it." According to him it will "outdo (the) A-6 by a wide margin" and use of the plane as a missileer to augment the F-14D in fleet air defense is "one of the concepts that's still viable." [11]

The A-12 program has provided a significant benefit to a legion of lawyers following the termination, beginning with a suit by the contractors to recover termination costs that the contracting officer had elected to deny. In December 1995, following a trial, the U.S. Court of Federal Claims issued an order converting the termination for default to one of convenience, which meant the contractors were entitled to termination costs. The Navy was successful on appeal in having the case remanded back to the Claims Court in July 1999, which after another trial in 2001 ruled that the contractors had been in default, which the contractors successfully appealed to the extent that the case was once again remanded to the Claims Court in 2003. (In 2002, the parties attempted to settle the case but could not agree on payments "in kind" by Boeing and General Dynamics that met the Navy's demand for repayment of $1.3 billion in progress payments plus $1 billion in interest.) In May 2007, the U.S. Court of Federal Claims once again upheld the default termination, which the contractors appealed. The appeal was still with the Court of Appeals of the District of Columbia as of September 2008.

Now What?

As of the beginning of 1991, the Navy's fighter and attack plans were in shambles. The A-6 upgrades had been canceled, the A-12 program had been terminated, the F-14D program was being targeted for termination by OSD, and the Navy was in the process of concluding that the NATF was unaffordable, not to mention another attempt to impose an Air Force-developed airplane on the Navy, a la the F-111 and NACF. Moreover, both the F-14 and A-6 were in dire need of

upgrading or replacement, with the major shortfall about to be the capability to perform an all-weather deep strike mission. In an ironic inversion of the late 1940s budget battles over strategic delivery of nuclear weapons, the Air Force was promoting its B-2 for worldwide overnight response to tactical situations in lieu of expensive carrier task forces. Once again, the relative cost and benefit of long-range bombers and aircraft carriers was being debated, only this time for conventional warfare.

Lacking stealthy airplanes, the Navy adapted the Israeli-developed Samson decoy as the ADM-141 Tactical Air Launched Decoy (TALD). Samson was deployed by the Israelis against Syrian missile sites in the Beka'a Valley in 1982 to misdirect and overwhelm the air-defense system. The Hornet can carry two 400-pound unpowered TALDs on one BRU-42 Improved TER. When dropped, it extends small wings and follows a preplanned profile, simulating an attacking airplane's radar signature (ADM-141A) or emitting chaff (ADM-141B). The Navy used it successfully in the first days of the 1991 Gulf War. The Iraqis, in trying to shoot down the decoys, wasted SAMs and exposed their radar sites to attack by HARMs. A subsequent improvement, the AMD-141C Advanced TALD, is powered by a small jet engine for increased flight time and a more realistic flight profile.

A bright spot for the Navy, if not Naval aviation, was the qualification and deployment of the Tomahawk surface-to-surface missile during the 1980s. The Tomahawk was an all-weather, long-range cruise missile that could penetrate enemy defenses at low altitude. It was a much smaller version of the surface-to-surface Regulus 1 missile, powered by a little turbofan engine, and launchable by booster rocket from non-aviation ships, including submarines. The land attack version initially navigated to the target by a combination of inertial navigation and terrain matching. The range was about 600 nm—roughly equivalent to the radius of action of the A-6—with a 1,000-pound warhead. (An anti-ship version equipped with targeting radar was developed and fielded; due in part to target-validation issues, all were subsequently converted to the land-attack configuration.) Its main drawback is the lack of reusability, particularly considering its expense, approximately $1,000,000.

The land-attack missile was first used operationally in Desert Storm in 1991. Subsequent improvements include the addition of GPS, the ability to be redirected after launch by satellite communication (including loitering while waiting for a target assignment), and longer range. The GPS-equipped version was first employed in Bosnia in 1995, significantly improving accuracy. It was employed in Operation Desert Strike in 1996, Operation Desert Fox in December 1998, Kosovo in April 1999, Afghanistan in 2001, and Operation Iraqi Freedom in 2003. Operational evaluation of the redirectable version, Tomahawk Block IV, was accomplished in late 2003 and it was deployed in 2004.

As effective as it was, Tomahawk couldn't fully replace the deep strike persistence and flexibility of a manned aircraft in locating and dispatching mobile targets as well as ones of opportunity, not to mention the value of the visible presence of a capability as opposed to the threat of one housed in the bowels of a ship offshore. The Navy still wanted to replace the A-6.

CHAPTER TWELVE

REPLACING THE A-6, ROUND TWO

After the cancellation of the A-12 program in January 1991, the Navy initiated the AX program to replace the deep strike capability of the A-6. The first-night mission requirement was an unrefueled radius of 700 nm—high-low-low-high—with four 1,000-pound bombs and two air-to-air missiles carried in weapons bays. It was to incorporate the latest technology in avionics, low observability, survivability, supportability, etc. The prospective contractors formed into five teams, with some contractors being on more than one team, the lead contractor of a team being listed first:

- Grumman/Lockheed/Boeing
- Rockwell/Lockheed
- McDonnell Douglas/LTV
- General Dynamics/McDonnell Douglas/Northrop[1]
- Lockheed/Boeing/General Dynamics

The program was expanded to include air-to-air capability and redesignated A/FX in recognition that the F-14 was also aging. Study contracts were provided to the five contractor teams at the end of 1991. The reports were due in September 1992. Based on the customary timetable for selection, development, qualification, and production, an airplane resulting from the new program wouldn't be available before 2006 at best.

In the meantime, the Navy had to be prepared to go into combat at any time, and the remaining A-6s were not going to last forever. Ever hopeful of new business, Lockheed management/marketers proposed carrier-based versions of the F-22 and F-117, but neither were of interest to the Navy compared to its own planes. The logical placeholders in the Navy portfolio were modifications and upgrades of the F-14 and F-18. The Navy's preference, the F-14, was at a disadvantage because Secretary of Defense Cheney had stated his determination to terminate the F-14D in 1989 along with reductions in other major programs.[2] There was no lack of F/A-18 variants to consider, in part because the original F/A-18 Hornet had disappointed in terms of payload and range. It was also limited in its ability to land back aboard with unexpended ordnance due to the landing weight limits. As the Navy transitioned to more expensive guided-weapons, this was becoming a bigger operational liability. The Navy and McDonnell's joint 1988 Hornet 2000 study, from which the Super Hornet evolved, addressed those issues as well as the need to reduce radar cross section and provide for growth in avionics weight and volume.

An F-14D and an F/A-18C face off against each other on the flight deck of Theodore Roosevelt in the Persian Gulf in January 2006. In fact, the battle for supremacy—there shall be only one—has already been decided. This is the F-14's last deployment. (U.S. Navy 060106-N-7241L-002)

The Lockheed/Boeing/General Dynamics team was promoting a variable sweep-wing variant of the F-22 for the NATF program when it was canceled. It was then considered as a candidate for the A/F-X program, which was to meet the Navy's need for both an A-6 and F-14 replacement. (Author's collection)

The F-14 was originally intended to have an air-to-ground capability. This early flight-test aircraft is the ultimate in strike fighter, taking off with 14 bombs in the tunnel between the engine nacelles, two drop tanks, two Sparrows, and two sidewinders. (Grumman History Center)

Both the F-14 and F-18 had their advocates. The F-14 was big enough to easily shoulder the A-6's bomb load and carry it a long way. Unlike most Navy fighters, however, it had not been qualified for air-to-surface weapons delivery even though it was initially intended to have a close air support and interdiction capability with up to 14,500 pounds of ordnance. The original software in the weapon system computer had included air-to-surface capability. Bomb drops were accomplished early in its development program, but initial problems and the qualification cost combined with the lack of need for additional bombing assets, given the presence of an A-6 squadron and two A-7 squadrons in the air wing at the time, resulted in the F-14 being assigned solely to the fleet air-defense mission.[3]

Grumman proposed strike upgrades to the F-14D, the least expensive of which was dubbed Quickstrike. It primarily involved avionics and software modifications to add air-to-ground weapons like Harpoon, HARM, and SLAM. More extensive airframe, engine, and avionics modifications, internally designated Super Attack Tomcat 21—for the 21st century—were also studied and promoted to the Navy as a low risk and cost-effective way of adding strike capability to the air wing.

The F-18 was smaller than the F-14, but it had the advantage of already being in service as a strike-fighter. It was also newer and more up-to-date from an avionics standpoint. It lacked long-range radar and Phoenix missile capability, the *raison d'être* of the F-14, but the "outer air battle" involving Soviet bombers equipped with cruise missiles had never occurred and it was increasingly unlikely that it would in the future. Last, but certainly not least, it was arguably less onerous to maintain than the F-14.

An F-14A/B Tomcat aircraft assigned to the Strike Aircraft Test Directorate of the Naval Air Warfare Center Aircraft Division Patuxent River is shown dropping 1,000-lb Paveway I series Laser Guided Bombs (LGB) on 1 February 1994 to clear the weapons for fleet use. Previous F-14 air-to-ground testing has been completed with Mark 80 series general-purpose bombs, cluster munitions, and various training stores. (U.S. Navy DN-SC-95-01058)

The F-14 air-to-ground capability was resurrected with the cancellation of the A-12 program. This NATC F-14D is performing a dramatic high-angle delivery of four 500-lb bombs with BSU-86 high-drag fins (sometimes called Snakeye IIs) dropped in the low-drag configuration as part of the weapons qualification. (U.S. Navy via author's collection)

Going forward after the A-12 cancellation, the Navy attempted to upgrade both the F-14 and F-18. VX-4 had already revisited the F-14 air-to-ground mode with Mk 83 and Mk 84 bomb drops in late 1987.[4] In July 1992, the F-14 was qualified to drop general-purpose bombs with clearance for CBUs and LGBs eventually following. McDonnell received a contract for the F-18E/F Super Hornet program in 1992. Although initiated by the Navy and approved by OSD as a modification program, it was a major change to what became known as the legacy Hornet.

The Navy was unable to convince Congress, much less OSD, to fund both the F-14D/Quickstrike and F-18E/F programs.[5] The decision in favor of the F-18 was ultimately decided on cost, although the analysis was flawed according to F-14 advocates. One argument was whether the much higher development cost of the F-18 Super Hornet was more than offset by its somewhat lower procurement and operating costs. Resolution was made more difficult by estimates that relied on different assumptions and data. The F-14 camp argued that the unit costs were based on significantly different annual production rates, 72 for the F-18 and 24 for the F-14. Basing the F-14 price on the same production rate as the F-18 reduced the F-14 premium to only 5 percent for, in their view, a more capable airplane. From the standpoint of operating costs, they complained to no avail that the F-14 data included the history of the F-14A, a much less maintainable airplane than the newer F-14D and less well supported in recent years.[6]

The Navy's independent A/FX program did not survive either. In accordance with the outcome of the Pentagon's 1993 Bottom-Up Review, it was canceled late that year and replaced with a joint service approach, the JAST (Joint Advanced Strike Technology) Program, which was to result in demonstrators of a supersonic strike fighter to replace the F-16, A-10, and legacy F-18s from 2010 to 2012. The Navy was to continue with the F-18E/F and the Air Force with the F-22. JAST was envisaged as two similar airframes with common avionics, engine, systems, and other components. More affordable and maintainable stealth was an essential requirement. A STOVL version was subsequently added to provide a replacement for USMC and Royal Navy Harriers as well.

Grumman F-14 Tomcat Upgrades and Proposals

In conjunction with the A-6F contract go-ahead in June 1984, Grumman had received another for an upgrade to the Tomcat, the F-14D. The major changes were a new engine, the GE F110, and a virtually all-new avionics suite, which included new multi-mode digital radar, the Hughes APG-71. The F-14D was finally the combination of engine, airframe, avionics, and weapons for fleet air defense that the Navy had hoped for when it initiated the program in the early 1970s. An F-14A modified to incorporate much of the F-14D avionics flew in November 1987. Three additional F-14As were also modified to test F-18D systems and features. To reduce cost and provide spares and support commonality, most of the mission avionics other than the radar—the mission computer, display processors, heads-up displays, and stores management system—were from the F-18A and were also incorporated in the A-6F.

The McDonnell/Northrop/BAE team would propose this JAST concept in the JSF competition. For VTOL flight, a lift engine was added to the forward fuselage in addition to thrust vectored from the main engine just before the afterburner stage. Another feature was the Pelikan tail, which was predicted to reduce weight, drag, and RCS by eliminating a separate vertical fin/rudder surface. (Boeing also considered using the concept and came to the conclusion that it resulted in a net weight increase.) (Author's collection)

However, Grumman management had reason to be concerned about competition for the air-defense fighter mission from a derivative of the Air Force ATF. Since the A-6F had been canceled, it was also in their best interest to resurrect the F-14's dormant strike capability and thereby increase its usefulness. There was also reason to believe that adding the air-to-surface mission would result in the Navy buying more F-14s and fewer, if any, A-12s as well as forestall F-18 upgrades.

Quickstrike: After its Northrop-led team was non-responsive to the contract terms of the 1987 competition that resulted in the A-12 and thereby "lost," Grumman began design studies of a strike-fighter variant of the F-14D. The resulting unsolicited proposal for a Quickstrike Tomcat (also called the "Block IV upgrade") was a minimal change to the F-14D to add FLIR and air-to-ground modes to its APG-71 radar, taking advantage of the software which already existed for the F-15E Strike Eagle's APG-70 radar. Inverse Synthetic Aperture capability would also be added, which the APG-70 lacked. Sea surface search and terrain avoidance modes were other enhancements.

In order to carry as many air-to-air weapons as possible, the F-14 already had stores attachment points on the belly between the two engine nacelles and on wing glove pylons. There were also attachment points on the nacelles for external fuel tanks. Removable, off-the-shelf pods mounted on the side of the wing glove pylons would provide all-weather navigation and targeting. One incorporated terrain-following radar and a wide-angle FLIR, and the other a steerable FLIR and laser

Tomcat 21 was Grumman's last best hope for continuing F-14 production and/or a major F-14 modification program. Unfortunately for Grumman, the Navy and Congress elected to invest in F/A-18 upgrades instead. (Tony Buttler)

target designator for targeting. Weapons qualification was to include the Standoff Land Attack Missile in addition to the Harpoon and HARM capability already planned for the F-14D.

The cockpit avionics would be changed to display the additional information from the radar and the pods, including a new heads-up display and a color moving map display. The cockpit would also become night vision goggle compatible. All of these changes were retrofittable to existing F-14s.

Attack Super Tomcat 21 was a variant of Super Tomcat 21. As proposed in 1988 in response to the formation of the NATF Program Office, it was to be a more significant modification to the engines, airframe, and avionics than Quickstrike, providing more range and more payload. The gross weight would be increased about 3,000 to 76,000 pounds. Mission radius with 8,000 pounds of ordnance for a high-low-low-high mission would be 550 nm, roughly matching the A-6's capability. Improvements included the higher thrust General Electric F110-429 engine; airframe modifications for a lower radar cross section from some aspects; bigger stabilators to maintain stability and control at lower approach speeds; more internal fuel in a thickened and widened wing glove area; modified flaps/slats for no-wind launch and significantly increased bring-back capability; and avionic upgrades to include the radar developed for the A-12 Avenger II program, an integral Night Owl FLIR/laser targeting system, and a helmet-mounted sight. Weapons-carrying capability would be added to pylon stations 2 and 7 under the nacelles that were only used for fuel tanks on the basic F-14s. None of the changes was so significant that existing F-14s with adequate airframe life remaining couldn't be rebuilt as Super Tomcats.

ASF-12 was Grumman's name for Advanced Strike F-14, the ultimate Tomcat, which it proposed and the Navy evaluated in 1994 after the AF-X program was canceled. Although not a clean-sheet-of-paper approach, it incorporated the highest possible technology available in addition to the advanced avionics from the ATA and ATF programs. For example, the engines would provide 3D-thrust vectoring. The outboard wing leading edge would have provisions for conformal radar. It was also to have a lower radar cross section.

"Bombcat" Briefly Fills the Gap

Congressional funding was approved at one point to start Quickstrike development, but the Navy had concluded in OSD-sponsored studies in 1990 and 1991 that the F-14D was not as survivable in the strike role as the F/A-18C/D and more expensive to procure and operate. The proposed F/A-18E/F was projected to be less expensive to operate and have a lower development risk than the more ambitious Attack Super Tomcat 21 or the ASF-14. The F-14's main selling point was its superiority in the defensive outer air battle because of its Phoenix missile armament. Unfortunately for Grumman, that capability was of diminished value because of the collapse of the Soviet Union and the availability of improvements in the surface-to-air defense provided by the cruiser-based AEGIS system.[7]

In any event, prospects for new production Tomcats with strike capability were severely handicapped by then Secretary of Defense Dick Cheney's decision in 1989 to terminate F-14D production with the last new aircraft to be delivered in 1992. Although contested, in the end only 37 new F-14Ds were built with another 18 produced by converting F-14As. Grumman was unable to convince Congress to override Cheney.[8] The last of the F-14Ds was delivered in November 1994. These 55 were used to equip three deployable squadrons, the Pacific Fleet training unit, and the test units such as VX-4.

Nevertheless, the cancellation of the A-12 in January 1991 and the steadily decreasing number of A-6s made the F-14 an appealing alternative for at least an interim deep-strike capability since it was capable of 600 nm radius missions, unrefueled. In 1990, the F-14A was finally qualified to deliver up to four Mk 84 dumb bombs on the fuselage pallet. Two Sparrows and two Sidewinders could be carried at the same time on the wing glove stations. In August 1990, VF-24 and VF-211 became the first fleet squadrons to qualify with this air-to-ground capability. In September 1995, VF-41 F-14s dropped laser-guided bombs (LGBs) as part of Operation Deliberate Force in Bosnia with designation provided by other aircraft.

In 1995, the Navy's Atlantic Fighter Wing initiated a program to provide a laser-designation capability to the Tomcat as quickly as possible. It bypassed much of the normal test unit development and qualification process by assigning responsibility to a fleet squadron, VF-103, at NAS Oceana, Virginia. Patuxent River was involved only for carrier suitability and electromagnetic compatibility tests. Lockheed Martin received a contract for a variant of its Low Altitude Navigation and Targeting Infra Red for Night (LANTIRN) system that had been developed for the USAF F-15E and F-16C. This was hung from the right hand multipurpose pylon. A hand controller and control panel were added to the aft cockpit with the FLIR imagery displayed on the existing tactical information display. An F-14B was modified to carry the pod and testing was completed by mid-June. For the initial operational evaluation, ten F-14s from VF-103 were modified and six pods were procured. Some of the Tomcats were also modified to have night-vision-goggle-compatible cockpits.[9] The squadron deployed aboard *Enterprise* in June 1996 and successfully demonstrated the utility of the system. The pods (and aircraft) were transferred to

This F-14D is returning to the Theodore Roosevelt *in October 2005 with unexpended ordnance, one each laser-guided and GPS-guided bomb, from a routine patrol. It is carrying the LANTIRN FLIR and targeting pod on the right glove station. This provided the crew with a day/night capability to accurately drop laser-designated bombs and fire the 20mm cannons.* (U.S. Navy 051023-N-5088T-002)

VF-32 in November so the evaluation and operational capability would be continued. VF-2 was the next to deploy with LANTIRN, the first on F-14Ds and also the first LANTIRN-equipped Pacific Fleet squadron.

The Bombcat was in place just in time to replace the A-6E payload and range performance, if not total mission capability, e.g., HARM, Harpoon, SLAM, etc. The last Intruder deployment, which was accomplished by VA-75—the first squadron to take the A-6A to sea—began in June 1996 aboard *Enterprise* and ended that December. The first F-18E had only just flown in December 1995. The first production airplane would not fly until December 1998. The first deployment of the single-seat version of the Super Hornet would not begin until 2002. Until then, the F-14s and legacy F-18s would have to be the main battery.

The F-14 squadrons went on to make major contributions to the carrier's strike performance. In 2000, VF-41 received the annual Clarence W. McClusky Award for air-to-ground strike warfare excellence, which was traditionally awarded to an attack squadron. The recognition was for 1,100 combat hours and 384 sorties over Kosovo during Operation Allied Force in 1999 while operating F-14As from *Theodore Roosevelt* (CVL-71).

After the successful operational employment of *Enterprise*-based VF-32's F-14s during Operation Desert Fox in December 1998, funding was authorized for qualification with 2,000-pound GPS-guided bombs for all-weather strikes, as well as to increase the LANTIRN designation altitude from 25,000 to 40,000 feet. The first JDAM drop in combat was made from a VF-11 F-14B in March 2002 on its

deployment with *John F. Kennedy*. F-14D JDAM qualification was completed and software modified in the field for its first combat drops in March 2003.

F-14As did not get JDAM capability. However, VF-154, operating F-14As as part of CVW-5 on *Kitty Hawk* during Operation Iraqi Freedom in 2003, dropped 358 laser-guided bombs in support of combat operations.

The last F-14 deployment was accomplished by VF-213 and VF-31 with Carrier Air Wing Eight aboard *Theodore Roosevelt* operating in the Persian Gulf and supporting ground forces in Iraq. Innovating to the end, the squadrons equipped their F-14Ds with Remotely Operated Video Enhanced Receivers (ROVER) data-transfer capability. This transmitted the image being recorded by the LANTIRN pod to a laptop receiver being viewed by a ground unit, providing it with a tactical picture that it could use for its operations or to coordinate with the Tomcat crew for weapons employment. The cruise ended in March 2006 and, after a few farewell performances, the U.S. Navy's Tomcats were grounded forever.

McDonnell F-18E/F Super Hornet

Everything about the Super Hornet was bigger than the legacy Hornet. It was 4 feet longer and had a 25 percent larger wing with thicker airfoils, less twist, and two additional stores stations. Internal fuel was increased by 33 percent. The LEX was enlarged and reshaped for better high angle-of-attack characteristics. As a result, structural commonality to the C/D was only 10 percent and the maximum gross

In 1996, F-14 fighter squadrons like VF-211 shown here were routinely training for air-to-ground weapons delivery in preparation for deployment. This F-14A is carrying four Mk 82 500-lb bombs in the tunnel. (U.S. Navy 960717-N-0226M-001)

Carrying something for everybody, this VF-102 F-14B is loaded with bombs, a Phoenix missile, and a Sidewinder. The long-range air-to-air Phoenix was unique to the F-14 and its radar system. It was originally intended to shoot down Russian bombers and cruise missiles attacking the carrier task force in the so-called outer air battle. At the moment, it's just another day in the Persian Gulf aboard George Washington in December 1997. (U.S. Navy 971202-N-2302H-004)

This VF-32 F-14B sports a tally of unguided and laser-guided bomb drops during Operation Iraqi Freedom while deployed with Harry S. Truman from December 2002 through May 2003. (U.S. Navy 030522-N-4953E-054)

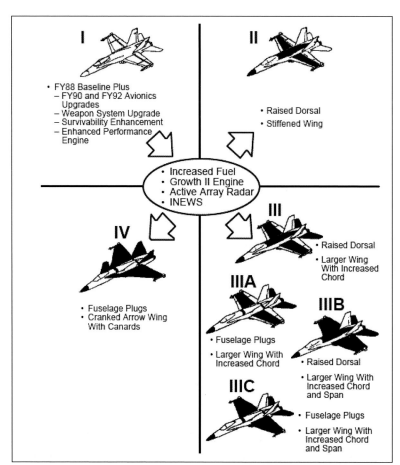

• I
  • FY88 Baseline Plus
    – FY90 and FY92 Avionics Upgrades
    – Weapon System Upgrade
    – Survivability Enhancement
    – Enhanced Performance Engine

• II
  • Raised Dorsal
  • Stiffened Wing

• Increased Fuel
• Growth II Engine
• Active Array Radar
• INEWS

III
  • Raised Dorsal
  • Larger Wing With Increased Chord

IV
  • Fuselage Plugs
  • Cranked Arrow Wing With Canards

IIIA
  • Fuselage Plugs
  • Larger Wing With Increased Chord

IIIB
  • Raised Dorsal
  • Larger Wing With Increased Chord and Span

IIIC
  • Fuselage Plugs
  • Larger Wing With Increased Chord and Span

McDonnell Douglas presented the Navy with several studies on how to restore or improve original range and endurance specified for the F/A-18. (INEWS stands for Integrated Electronic Warfare System.) The F/A-18E resembles option IIIC. (Author's collection)

Improvement (P^3I) or "spiral development" of the avionics. The Super Hornet suite would be periodically upgraded to add functionality and capability as warranted by the availability of proven advancements in technology. This was to result in a Block 2 F/A-18E/F.

Numerous detail changes were introduced as well. The mechanical backup control system was deleted. The speed brake was deleted; when needed, more drag was provided by the flaps going down, the ailerons and spoilers going up, and the rudders both deflecting out while the stabilators maintained the pitch attitude. Spoilers and vents were added to the leading-edge extensions, the former primarily to assist as a speed brake and the latter to replace the fences deleted for radar signature reduction. (The vents were deleted after flight-test evaluation showed them to be ineffective.) Components were redesigned or relocated to reduce vulnerability and a fire-suppression system was added.

An increase in landing weight allowed the pilot to land back aboard the carrier with 9,000 pounds of fuel and ordnance compared to the 5,500 pounds for the legacy Hornet. That enabled it to return with a 2,000-pound GBU-24 bomb, a capability that the legacy Hornets did not have if 4,000 pounds was the required minimum fuel to have on its first approach. F/A-18C/Ds were not usually assigned GBU-24 missions because of that restriction.

The F/A-18E was to be a single-seat airplane like the F/A-18A/C. The F/A-18F was to have two seats, with the rear cockpit configured for a Weapons System Operator like the F/A-18D. Compared to the F-14, the F/A-18F not only lacked the previously all-important Phoenix missile system, its radar had half the detection range of the F-14's. It also did not match the F-14 in payload and range:

	F/A-18F	F-14 Quickstrike
Payload	4 x Mk 83	4 x Mk 83
	2 x AIM-9	2 x AIM-9
External Fuel (gal)	2 x 330	2 x 280
Mission Radius (nm)	400	460

The F/A-18F's range shortfall increased with heavier bomb loads. When the bomb load was doubled, it had a mission radius of only 265 nm compared to 400 nm for the F-14. The Tomcat was also faster than the Super Hornet. However, after the investment in development and qualification had been made, it was arguably cheaper to buy and operate.

OSD approved the F/A-18E/F program in May 1992 for engineering and manufacturing development, waiving the prototyping phase because it was presented as a low-risk modification of the existing Hornet. In July, the Navy awarded a contract to McDonnell Douglas. While strongly supported by Navy leadership, the Super Hornet, like its predecessor, nevertheless still had critics within the service and the retired community, not to mention Congress.

The first Super Hornet, a single-seat E, flew in November 1995, a month ahead of schedule, closely followed by the second test aircraft, another E. Flight test commenced at Pax River in December. One of the early disappointments uncovered in March 1996 was that one wing would abruptly lose lift in high-speed, high-g maneuvers. The Super Hornet would roll out of a hard turn, uncommanded by the pilot.

weight had increased to 66,000 pounds. The engine was the General Electric F414, based on the F412 intended for the canceled A-12 and the F-18's F404, with 36 percent more thrust than the F404. It carried bigger external tanks, 480 gallons versus 330.

The design was also modified to reduce the radar cross section from critical aspects. The most obvious was the reshaping of the larger engine inlet ducts to be parallelograms. Inside the inlets, a mechanical device blocks the radar return of the engine compressor. Low observable materials were also incorporated in critical areas, and landing gear door and access panel edges were aligned to minimize radar reflection. The radar cross section was reportedly 90 percent less than that of the smaller C/D but not in the same league as the Air Force's ATF, now the F-22.[10]

Also unlike the F-22, it didn't have supercruise, thrust vectoring, internal weapons bays, or an all-new avionics suite with Advanced Electronically Scanned Array (AESA) radar.[11] The relative lack of advanced technology, however, meant that it would be delivered sooner at less cost. The Navy also had a plan for Preplanned Product

The wing-drop problem was *déjà vu*. There had been development problems with the original F/A-18 wing, which had been resolved in part by the deletion of the outboard wing leading-edge extension that McDonnell had added to the Northrop design.[12] As one of the many changes to the legacy Hornet design, the snag had been reintroduced on the F/A-18E/F's wing in anticipation of an aerodynamic benefit. Once again, there were unanticipated consequences. Attempts to correct it with software changes to the control system weren't successful. Configuration fixes were evaluated with a solution finally qualified in 1998: a porous skin over the wing fold area which minimized the differential loss of lift between the left and right wing at high angles of attack.

The Super Hornet continued to be criticized by the GAO and the CBO as being too expensive, not meeting performance projections, and in some characteristics, like acceleration and maneuverability, inferior to the F/A-18C/D. One option, which was discussed in Congress, was to extend the life of the F-14s until the Joint Strike Fighter (JSF), the product of the JAST program, was available. Notwithstanding the frequent expression of concern and even outrage, the program continued to be funded. It helped that with the exception of the wing drop problem, development was relatively trouble free and on schedule, within cost, and most remarkably, under the specification weight.

Initial at-sea carrier qualification trials were accomplished aboard *John C. Stennis* (CVN-74), with the first F/A-18F in January 1997. Approach speeds were about 10 knots slower than the F-18C/D. The projected average unit cost was also reduced by a 140-unit increase in the total quantity planned as a result of the Navy's decision to replace the EA-6B with an electronic warfare variant of the F/A-18F.

Formal operational evaluation began in May 1999 and resulted in the Super Hornet being declared "operationally effective and operationally suitable" without the histrionics experienced by its predecessor in the same phase. By military procurement standards, development was completed on cost and schedule. The flying portion of the operational evaluation was completed in November 1999, the

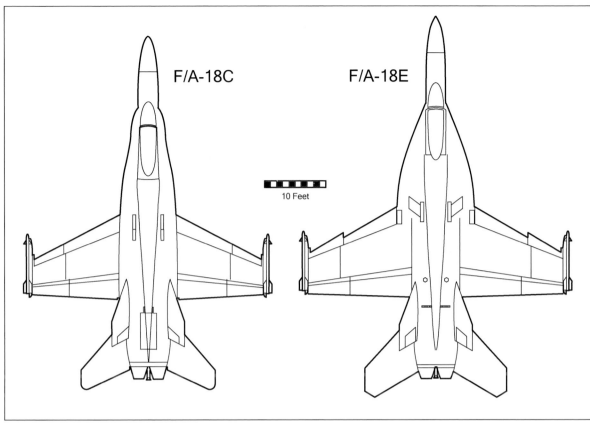

The major external differences between the F/A-18C and the F/A-18E were a stretch in the fuselage, an enlarged leading edge extension, and larger wing area. The differences were more than skin deep. It was all but entirely new and what wasn't, like some avionics, was planned for upgrade in the near future. (Author)

same month that the fleet-readiness training squadron, VFA-122, received its first airplanes. VFA-115 was the first operational F/A-18E squadron. It deployed aboard *Abraham Lincoln* (CVN-72) in July 2002, just in time for combat missions in support of Operation Enduring Freedom in Afghanistan and Operation Iraqi Freedom. The two-seat F/A-18F first deployed with the third Super Hornet squadron, VFA-41, and saw combat for the first time in April 2003.

Another collateral duty for the Super Hornet was in-flight refueling. Four of the wing pylons and the centerline station were "wet," enabling it to carry four external tanks and the refueling pod. With the retirement of the Lockheed S-3 (its last deployment was aboard *Enterprise* and ended in December 2007), another type needed to be substituted in this all-important supporting role, particularly since as some wags would have it, one of the first radio calls from a legacy Hornet pilot after takeoff was the request "Steer for Texaco," meaning "Where is the airborne tanker?" Unlike the KA-6D, which was dedicated to the tanking mission with integral hose and drogue equipment and had no attack radar, any F-18E/F can be configured to be an in-flight fuel provider.

One of the defensive innovations introduced with the F/A-18E/F was the ALE-50 Towed Decoy. It was deployed from a launcher

The first F/A-18E/F was used for initial at-sea carrier qualification in January 1997 aboard John C. Stennis. *Note the larger and rectangular engine inlets, the former for the greater mass flow required by the more powerful engine and the latter for forward-aspect radar cross-section reduction.* (U.S. Navy 970118-N-4787P-002)

mounted just behind the centerline pylon and provided, in the words of the manufacturer, "a preferential target that lures enemy missiles away by providing a much larger radar cross section than the aircraft." The ALE-55 Fiber Optic Towed Decoy (FOTD) was an aerodynamic improvement to the decoy to minimize stress on the line, the use of a fiber optic connection to aircraft's countermeasures avionics for improved response to the threat radar, and a strengthening of the tow line so that it was less likely to burn through when dragged into the engine exhaust plume. Three decoys are carried per launcher, with a deployed decoy being severed before landing.

The Navy F-117

The Navy kept itself informed about the F-117 program, including having Naval aviators fly it during late 1984 and assigned to the operational Air Force squadrons as exchange pilots. From time to time, Lockheed provided carrier-capable design studies to the Navy that elicited little interest. However, in early 1993 the Lockheed Skunk Works began to market a fully navalized and upgraded F-117 as an A/FX candidate.

The basic changes were relatively minor: the addition of a catapult-compatible nose gear, higher sink rate-capable landing gear, a tailhook, a new wing that was less swept and could be folded, and a horizontal tail. The basic F-117 structure was amenable to these changes because it had a full-depth center keel from the nose gear to the tailhook, three full-depth fuselage frames for wing carry through, and a main landing gear mounted directly to a major bulkhead. With the wings folded, its spot factor was bigger than the A-6's but less than the A-12's.

The wing planform change, combined with trailing-edge flaps and leading-edge slats, and the addition of the horizontal tail provided the necessary reduction in minimum flying speed for carrier takeoffs and landings. Spoilers were incorporated to provide improved roll control at low speeds and direct lift and on approach. Speed brakes were added on the fuselage to improve maneuvering capability. On approach, extended speed brakes also resulted in a higher engine RPM for quicker thrust response in the event of a wave-off.

The bomb bay was enlarged by increasing the keel depth by 19 inches and bulging the bomb bay doors. This doubled the internal payload capacity to 10,000 pounds. Air-to-air missiles could be mounted on the inside of the doors. Two removable stores pylons were also added under

Unlike the legacy Hornets, the Super Hornets could be launched with enough fuel to function as a tanker. Here a VFA-22 F/A-18E is refueling a VAQ-139 EA-6B also deployed on Ronald Reagan in mid-2006. The tanker configuration consisted of four external tanks and a refueling store. (U.S. Navy 060701-N-4776G-028)

Because of its higher margin of landing weight to empty weight, this F/A-18E will be easily able to bring back the JDAM that it is taking off with from Abraham Lincoln in October 2002. The same could not be said for the legacy Hornets. (U.S. Navy *021004-N-9593M-038*)

each wing so that an additional 8,000 pounds of ordnance or fuel could be carried when full stealth was not required.

With 10 knots WOD, maximum launch weight was 68,750 pounds compared to the F-117A's gross weight of 52,500 pounds. Mission radius, unrefueled, was 700 nm with 6,600 pounds of ordnance. Lockheed also projected that the aircraft could land back aboard with even more fuel and ordnance, 11,600 pounds, than the F/A-18E/F.

The engines were afterburning General Electric F414s, the same as the F/A-18E/F. The avionics suite incorporated a radar, which the basic F-117 did not have, with all-weather air-to-ground and air-to-air capability. The latter, along with a canopy with fewer frames and provisions for air-to-air missiles, provided a multi-mission capability, which the basic F-117 lacked. Lockheed claimed that it would have approximately the same specific excess energy, a measure of maneuverability, as the F-14D.

It was not an easy sell, even though it was arguably a next-generation strike fighter with an order of magnitude better stealth than the F/A-18E/F. Detractors criticized it as 1970s technology. RAM maintenance requirements had originally been a challenge and at the level initially experienced by the Air Force, unacceptable in a carrier-based environment. However, low observable improvements had been designed and evaluated, which promised to significantly reduce the maintenance effort required while retaining, if not improving, effectiveness. However, that was hard for anyone not knowledgeable about stealth technology to judge and, moreover, the benefits had not yet been proven at the time.

The F/A-18E/F was also designed with a reduced radar cross section, particularly from the forward aspect. This—combined with accompanying electronic warfare and SAM-site suppression aircraft,

The AGM-154 Joint Strike Standoff Weapon (JSOW), shown here in red on an F/A-18E flight test, was designed to replace the Cluster Bomb Unit with a delivery capability that did not require a close-in attack. The Navy also developed a variant as a Walleye replacement. The JSOW has extendable wings that provide a standoff range of 15 nm for a low-level release and 40 nm for a high-altitude release. Both INS and GPS navigation are provided. The Walleye replacement has a thermal imaging seeker and warhead for penetration of hardened targets. (Terry Panopalis collection)

This 1992 drawing of the F-117N Seahawk by Steven Moore depicts the basic planform changes and the fighter-type canopy but not all details like wing fold or the leading edge flaps. At this point, the airplane retains most of the basic fuselage structure except for the canopy and engine exhausts. The engine inlets also appear to be larger. (Jay Miller collection)

The A/F-117X had a deeper bomb bay to accommodate greater bomb loads internally. As shown in this 1995 drawing by Steven Justice, air-to-air missiles could be mounted on the interior of the bomb bay doors for self-defense. (Jay Miller collection)

This 1993 artist's concept by Steve Justice of a Lockheed "Seahawk" about to touch down depicts the flap and spoiler system and the speed brake on the side of the fuselage. F-14/A-6-type nose and main landing gear were to be used for carrier operations. At this point, the design still incorporates the F-117 tail fins and the sharply angled wing fold line. (Jay Miller collection)

self-protection devices, the forthcoming AESA radar, and other systems/tactics—at least partially addressed the survivability concerns that were more fully answered by an aircraft as stealthy as the F-117.

Nevertheless, there was considerable support in the Navy operational leadership for this version of the F-117 and even more importantly, in Congress. It would have provided a first-day strike capability that had been ceded to the Air Force in Desert Storm. However, there wasn't enough money in the budget to do both the Super Hornet and the F-117. To spread the development cost and reduce the average unit cost, Lockheed also tried to include the Air Force in this new buy by proposing an "F-117B" with the same improvements, other than those for carrier basing. The reduction in empty weight and greater land-based takeoff weight would have resulted in even greater unrefueled radius of action, almost 1,000 miles with full internal ordnance.

The Senate Armed Services Committee expressed its support by directing that $175 million of OSD's FY1996 requested budget be used to initiate a program definition phase and build a flying demonstrator of a new production aircraft. The resulting Defense Authorization bill didn't go that far but did include $25 million for a six-month A/F-117X program definition, which NavAir implemented.

In the end, the Navy decided that it couldn't take the risk of forgoing or even delaying the F/A-18E/F that had just flown and switch to a much stealthier airplane that might be unsatisfactory from a maintenance standpoint, given how few aircraft an aircraft carrier could carry in total. Although Lockheed undoubtedly claimed otherwise and had the reputation to back it up, the switch also risked a delay in the availability to the fleet of an A-6/Bombcat replacement, which was already overdue.

If the Air Force had opted for the F-117B and the Navy's unique portion was affordable, it would probably have been deemed an

The definitive A/F-117X—as proposed in 1995 and drawn here by Pruitt Benson—featured a redesign of the vertical fins to shorten the aircraft. Fixed fins and rudders have been substituted for the all-moving fins and the overall area is larger to compensate for the shorter moment arm. The horizontal stabilators have been increased in area. The wing fold hinge line now runs more or less straight fore and aft. (Jay Miller collection)

acceptable risk, resulting in at least the existence of a deployable squadron of Super Sea Hawks that would fly aboard whenever its unique capability was needed. Unfortunately for that scenario, the Air Force was an even harder sell since it was investing heavily in the stealthy F-22. The Navy would have to make do with the F/A-18E/F for first-day strike missions until the next generation of stealthy strike fighters entered service.

JAST X-32 and X-35 Demonstrators

The JAST program—initiated in 1993 to replace the Air Force A-10 and F-16 and the Navy/Marine Corps F/A-18A-D and the AV-8B—begat the Joint Strike Fighter (JSF) program in late 1995. Stealth was one of the design requirements. After a paper competition that included a McDonnell Douglas/Northrop team, Boeing and Lockheed each received contracts in November 1996 to build the X-32 and X-35 demonstrators, respectively. Only two prototypes were to be built per contract, but Boeing and Lockheed each had to provide three different configurations for evaluation: conventional land-based, conventional carrier-based, and short takeoff/vertical land (STOVL). The highest degree of difficulty was clearly the STOVL version. Boeing and Lockheed had chosen two completely different approaches. Boeing's concept was similar to the Harrier, but instead of vertical thrust being provided by four swiveling nozzles, there were only two diverter nozzles located at the aircraft's center of gravity.[13] Lockheed's STOVL concept was much more risky technically: a single swiveling nozzle at the rear of the airplane with an engine exhaust-driven turbine mechanically driving a counter-rotating lift fan located just aft of the cockpit. The fan drive was de-clutched for up-and-away flight. In addition to providing lift, the fan thrust was also varied for hover control in pitch. Part of the high degree of difficulty was the enormous amount of horsepower—28,000—that the clutch had to withstand and the fan gearbox had to turn through 90 degrees.

Lockheed's fan was high risk because of the gearbox, clutch, and shafting involved, but it also provided much more efficient lift in hover than jet thrust could. Moreover, when the fan was removed, the vacated volume in the same basic airframe was ideally located for the addition of more fuel and/or avionics or a second cockpit to a conventional takeoff and land airplane. Fortunately for Lockheed, the gamble paid off when Allison Engine Company was able, just in time, to successfully develop and qualify the fan. The X-35's hover performance was significantly better than the X-32's and contributed in no small part to the selection of Lockheed in October 2001 for the system development and demonstration of the F-35.

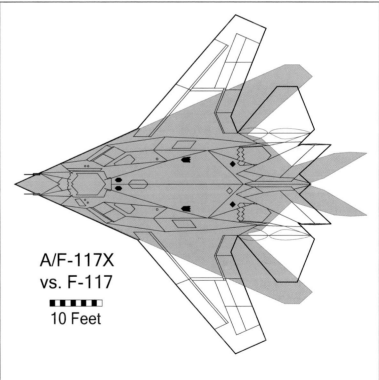

A/F-117X
vs. F-117

■■■■■
10 Feet

In order to provide an acceptable speed for carrier-based operations, the F/A-117X had a horizontal tail, less wing sweep, and more wing area along with leading-edge slats and trailing-edge flaps. Less and less of the basic F-117 fuselage structure was retained as the aircraft was optimized for the Navy's mission requirement. (Author)

The two contenders for the JSF program, the Boeing X-32 on the left and the Lockheed Martin X-35 on the right, are posed together on the ramp at Edwards Air Force Base. The Boeing has the more innovative aerodynamic configuration. The Lockheed vertical takeoff and landing propulsion concept incorporating a lift fan is high risk but potentially provides more thrust in VTOL mode than simply diverting the engine exhaust. (Author's collection)

The naval variant of the Lockheed Martin JSF, the X-35C, was only used for handling quality and performance testing, not actual arrested landings. The major external difference besides the tailhook was a larger wing. It is shown here arriving at NAS Patuxent River in February 2001 for evaluation. (U.S. Navy 010210-N-0000P-001)

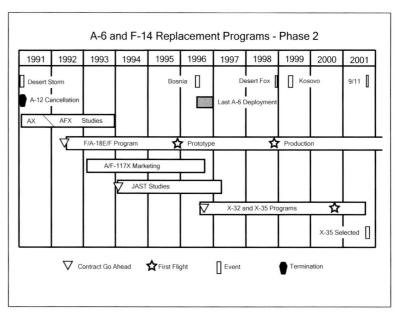

By the end of 2001, the McDonnell was well on its way to achieving total domination in carrier-based strike aircraft by the end of the decade with the F/A-18. However, the next generation strike fighter, the Lockheed F-35, was about to enter full-scale development to supplant if not replace it. (Author)

CHAPTER THIRTEEN

SUMMARY

With the retirement of the F-14, the number of fixed-wing aircraft types in an air wing was reduced to four, shown here in formation above Kitty Hawk in June 2008. They are (from left to right) the E-2C Hawkeye, the F/A-18C, the F-18E/F, and the EA-6B. In a few years, it will be three, with the EA-6B replaced by the EA-18G. (U.S. Navy 080623-N-7883G-274)

The U.S. Navy's carrier-based attack community has experienced major changes in the last 60-plus years and is considering yet another: routine operation of unmanned aircraft. Many of the innovations were foreshadowed by World War II projects: jet aircraft, electronic warfare, all-weather attack solely by radar, remotely piloted or autonomous missiles, etc. Some technologies could only have been dreamed of then—digital computers, inertial navigation, laser designation, extremely accurate navigation using signals from satellites, multifunction displays, and night vision.

One constant is the aircraft carrier, a mobile and sovereign air base operating tactical aircraft that are the equal of any land-based ones they will face. Another is the relentless pursuit of greater survivability in the face of more effective defenses while still achieving better weapons delivery accuracy. Closing to pointblank ranges to deliver bombs and torpedoes as practiced in World War II was no longer viable. As a result, standoff weapons were developed, and the aircraft-delivered torpedo was abandoned as a weapon to be used to sink surface ships, although it still had a role in antisubmarine warfare.

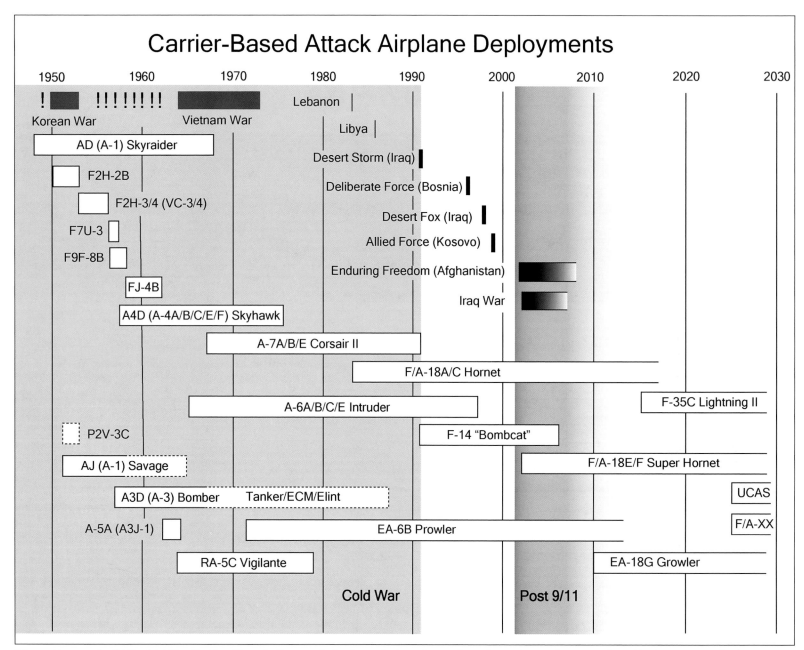

Carrier-Based Attack Airplane Deployments

| Timeline | 1950 | 1960 | 1970 | 1980 | 1990 | 2000 | 2010 | 2020 | 2030 |

Korean War
Vietnam War
Lebanon
Libya
Desert Storm (Iraq)
Deliberate Force (Bosnia)
Desert Fox (Iraq)
Allied Force (Kosovo)
Enduring Freedom (Afghanistan)
Iraq War

AD (A-1) Skyraider
F2H-2B
F2H-3/4 (VC-3/4)
F7U-3
F9F-8B
FJ-4B
A4D (A-4A/B/C/E/F) Skyhawk
A-7A/B/E Corsair II
F/A-18A/C Hornet
A-6A/B/C/E Intruder
F-35C Lightning II
P2V-3C
F-14 "Bombcat"
AJ (A-1) Savage
F/A-18E/F Super Hornet
A3D (A-3) Bomber   Tanker/ECM/Elint
UCAS
A-5A (A3J-1)
EA-6B Prowler
F/A-XX
RA-5C Vigilante
EA-18G Growler

Cold War   Post 9/11

Since the end of World War II, the Navy's forward-deployed aircraft carriers have been frequently called on to actively protect America's interests abroad or signal the willingness to do so by presence. The strike aircraft on board and the weapons available have been continuously improved to maintain the credibility and effectiveness of this national asset. In the 1950s and early 1960s, the attack squadrons transitioned from propeller-drive to jet airplanes, with various jet fighters filling in for the forthcoming A4D Skyhawk. (The exclamation points denote the mini-crises that the aircraft carriers were ordered to respond to.) The development of heavy attack airplanes ended with the deployment of the Polaris missile but resulted in the most capable reconnaissance airplane, the RA-5C, which the Navy was to employ. By the end of the 1970s, all strike aircraft were all-weather capable and employed sophisticated defensive electronic suites. The next two decades saw the introduction of air-to-ground guided weapons of remarkable accuracy. Beginning in the mid-1990s, unrefueled strike radius was diminished with the retirement of the A-6 and the F-14 "Bombcat." The non-aviation ship-based Tomahawk missile offset the loss of range, but it is otherwise not a replacement for the breadth, depth, and flexibility represented by manned aircraft. In the 2000s, the Navy is continuing to fund technology for, and development of, yet another generation of carrier-based strike aircraft and weapons. (Author)

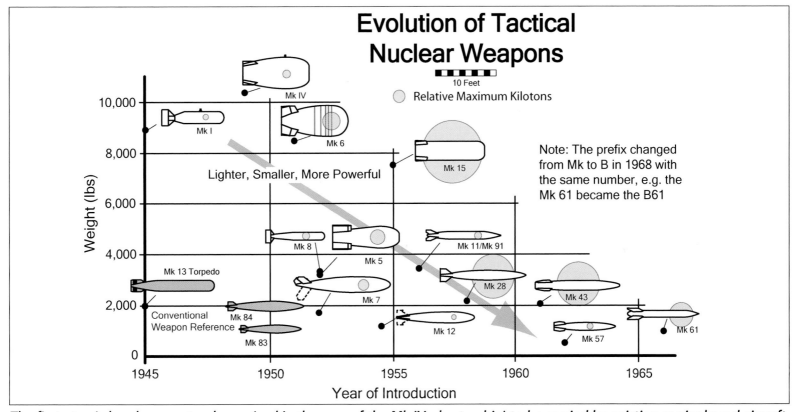

Evolution of Tactical Nuclear Weapons

10 Feet

⬤ Relative Maximum Kilotons

Mk IV

Mk I

Mk 6

Mk 15

Note: The prefix changed from Mk to B in 1968 with the same number, e.g. the Mk 61 became the B61

Lighter, Smaller, More Powerful

Mk 8

Mk 5

Mk 11/Mk 91

Mk 13 Torpedo

Mk 28

Mk 43

Mk 7

Conventional Weapon Reference

Mk 84

Mk 12

Mk 57

Mk 61

Mk 83

Weight (lbs)

10,000 — 8,000 — 6,000 — 4,000 — 2,000 — 0

1945   1950   1955   1960   1965

Year of Introduction

The first atomic bombs were too heavy (and in the case of the Mk IV, also too big) to be carried by existing carrier-based aircraft. This led to the development and deployment of the AJ Savage and A3D Skywarrior. In the early 1950s, the Mk 7 and 8 were qualified with yields similar to the first atomic bombs and yet small enough in size and weight to be carried by tactical airplanes, although the Mk 7 was very bulky and the Mk 8 still heavy compared to conventional weapons. By 1955, the Mk 12 replaced the Mk 7 and was comparable in size and weight to the Mk 83 and 84 high-explosive bomb. The Mk 57 only weighed 500 pounds and had the same yield as the Mk 3. The Mk 8's replacement, the Mk 11, was still heavy because it was a gun-type weapon designed to destroy deeply buried bunkers and well-sheltered submarine pens. The Mk 15 had a yield of two to three megatons—it was the first "lightweight" thermonuclear bomb. Further development resulted in the Mk 28 and the Mk 43, which were equivalent in weight to the Mk 7 but had yields of up to one megaton. (Author)

One significant change is that there are no longer any dedicated attack aircraft in the air wings. Budget realities have resulted in the transfer of the Navy's deep-strike infrastructure destruction mission to Tomahawk land-attack missiles launched from submarines and surface ships. Dual-mission airplanes now accomplish the close air support and other air-to-ground missions, as much fighter as attack in capability.

The first major crisis involving the Navy attack community immediately after the war was the need to develop big, long-range nuclear bombers. Successfully accomplished, that era ultimately ended in the mid-1960s with the deployment of the Polaris missile submarine, a far more effective solution than manned aircraft in the event that a nuclear exchange was unavoidable. (Some believe that the big bombers became superfluous long before that, when smaller nuclear bombs and inflight refueling were introduced as described in Chapter 6.)

The remainder of the attack community continued to be tasked with special weapons delivery as a high priority. The first carrier-based jet designed specifically for light attack, the A4D, was optimized for the nuclear delivery mission. Fortunately, it proved adaptable to conventional warfare because that was what it was called on for at the beginning of the Vietnam War. Follow-on jet attack airplanes continued to be tasked with the nuclear delivery mission until September 1991, when President George H. W. Bush directed that tactical nuclear weapons be removed from U.S. surface ships, attack submarines, and naval aircraft. This was reported complete on 2 July 1992.

With the exception of the A3J Vigilante, the Navy was steadfastly subsonic with respect to attack airplane performance until its development and procurement plans began to be shaped more by OSD and Congress than its own experience and expertise. The resulting airplane, the F/A-18, was not only supersonic, but a derivative of an Air Force technology demonstration program. Even though unwanted, disliked, and ridiculed by many within the Navy during development and initial deployment, it not only survived but also prevailed over all the periodically proposed alternatives.

The AM Mauler cockpit represents the past—a specific gage for each indication and a specific switch or lever for every action. (The gunsight and standby compass are missing.) Note the radar scope on the upper right side of the instrument panel. (Lockheed Martin *Code One*)

The F-35 cockpit represents the present—large multi-function displays, fewer switches, and data managed by computer. Many of the most necessary switches are provided on the stick and throttle. Note the lack of a heads-up display. The pilot is now wearing it. (Tom Harvey)

Both the Air Force and Navy developed weapons enabling attackers to stand off farther for increased survivability. In most cases, however, the initial advantage was all but nullified by innovations in surface-to-air weapons. A new imperative for strike mission effectiveness had also already become evident by the end of World War II: electronic warfare. The fundamentals—denial of enemy radar and communications capability for air defense—did not change significantly, but the degree of difficulty and penalty for failure increased exponentially with the development of surface-to-air missiles. An aircraft dedicated to the protection of the strike group proved to be required. The threat also affected strike aircraft design, dictating an emphasis on self-protection from radar and infrared guided missiles and at least some consideration of radar cross section and infrared signature reduction.

As important as not being hit was the capacity to take a hit and still return to base. Often overlooked or "traded away" to achieve empty weight and then added back when combat losses exposed the weakness, vulnerability reduction became a mandatory element of aircraft systems engineering, equal to weight in importance. "Live fire" testing was eventually a specific qualification requirement to prove that the vulnerable area was as small as specified and implementation of technology like fire and explosion suppression in fuel tanks, effective. Although vulnerability reduction is usually achieved by duplicating or protecting vital systems, another approach was deleting them. Ed Heinemann's effort to minimize the A4D Skyhawk's weight by eliminating as much as possible was partly responsible for its good reputation for survivability in Vietnam.

The need for all attack aircraft to operate in all weather conditions and at night drove cockpit design and the operation of systems like radar that had previously been reserved for specially trained and qualified aviators. This all-weather capability initially resided in large shore-based units that provided detachments to air groups. Today, the single-seat F/A-18 pilot has displays and controls that provide a degree of situational awareness and systems management undreamed of by an earlier generation of pilots and not only operates in all-weather conditions but is capable of performing both fighter and attack missions.

An example is navigation. Light attack jet pilots in the 1950s initially found their way just like Charles Lindbergh had between New York and Paris, flying a heading at a known speed, both corrected for estimated wind direction and velocity, for an elapsed time, updating the

The latest fashion in pilot's headgear includes a visor with selectable displays, including day and night versions, which replaces the heads-up display. Head tracking is used to change what is displayed in azimuth and elevation for better situational awareness. For example, if the pilot looks down with infrared imaging selected, he will in effect be looking through the cockpit floor. (Lockheed Martin *Code One* Magazine)

Representing the sharp contrast between today's head gear and navigation displays and yesterday's, Ensign Robert Bennett is pictured on 1 May 1951 just before launching in his Skyraider to destroy a lock of the Hwachon Dam in Korea with a World War II surplus Mk 13 torpedo. The only special feature of his goggles is the sunglare-reducing tint. He's carrying the chartboard that slides into the instrument panel, his navigation charts, and his knee board on which he records checkpoint times and fuel usage. (U.S. Navy photo via Robert L. Lawson collection)

track periodically by the visual observation of prominent landmarks. The crews of big all-weather bombers were somewhat better off because they had radar and could also take star sights and determine position just as Columbus did. The introduction of inertial navigation systems, moving map displays, and lightweight radars in the 1960s provided the single-seat jet pilots with the tools to fly safely at night and in all weather conditions. The Global Positioning System introduced in the 1980s allowed any pilot to use "stars"—without even looking up, much less at tables—for increased position accuracy and simplified updating the INS reported position, which drifted over time.

Weapons-delivery accuracy was initially focused on improving the pilot's aim and release so that some of the unguided weapons would hit near enough to the target to damage it. The practical limit in unguided accuracy was reached with the A-7D/E and its Continuously Computed Impact Point.[1] However, the CEP achieved with unguided weapons still required several bombs to destroy a target and in combat was degraded by the greater standoff distances necessitated by improved defenses. TV-guided or

Air-to-surface weapon development during and between wars emphasized increases in accuracy and standoff distance. Although laser-guided bombs were available in the early 1970s and evaluated in combat in Vietnam, the Navy was very slow to deploy them, unlike the Air Force, and dropped few in Desert Storm in 1992. By the end of the 1990s, however, precision-guided weapons were the rule and dumb bombs the exception for carrier-based aircraft missions. (Author)

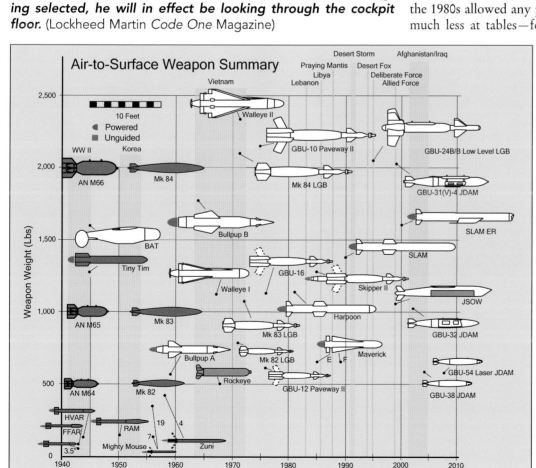

Air-to-Surface Weapon Summary

Carrier-based deep strike was effectively ceded to non-aviation ships after the failure to replace the A-6 Intruder in terms of payload/range performance. Here a Tomahawk cruise missle is being launched from the guided missile cruiser Cape St. George (CG-71) in March 2003 from an undisclosed location in the Mediterranean Sea. It has already begun to turn on course toward its target. (U.S. Navy 030323-N-6946M-002)

This is the finale of a test of the Block IV Tactical Tomahawk in August 2002, about to destroy the traffic cone target. This upgrade provides in-flight retargeting, battle damage assessment capability, and in-flight health and status reporting through a satellite data link. In other words, it is beginning to approach the mission flexibility of the manned aircraft. (U.S. Navy 020823-N-9999X-002)

flare-aimed weapons were expensive, involved relatively high workload, and were merely more accurate than unguided bombs as opposed to precise. The answer was first laser guidance and then GPS guidance. While not quite achieving the one target, one bomb metric promoted by the military services and contractors, the precision was to a large degree independent of standoff range, improving survivability and resulting in far fewer sorties required to accomplish the mission objective. Early laser-guided bombs finally brought down the North Vietnam's Thanh Hoa Bridge in 1972 after seven years of strikes with tons of bombs, missiles, and mines that had minimal effect. Up until then, at least 10 aircraft had been shot down of the several hundred that attacked the bridge or its defenses beginning in 1965.

Increased accuracy benefited carrier-based forces even more than those land-based because of the physical size limitation for bomb storage on the ship. The unit cost was also significantly less than previous precision weapons. To increase tasking flexibility, the Navy is updating both GPS-guided JDAMs and laser-guided bombs for dual-mode capability.

Strike U

Achieving results and minimizing losses in a strike takes more than just state-of-the-art airplanes and weapons systems. After the United States withdrew from Vietnam in 1975, the carrier-based Navy was not called upon to conduct an opposed strike until December 1983, when two F-14s flying over Lebanon were fired upon by Syrian antiaircraft artillery. A retaliatory strike was launched the next day from *Independence* and *Kennedy* on very short notice. For some rea-

son, the plan that the two Air Wings had devised and were implementing was countermanded by orders from Washington changing the weapons plan and moving up the time-on-target. The result was a scramble to brief, reconfigure the weapons loading, and launch the strike on the new schedule, heading into the rising sun. One outcome of the confusion was the late launch of EA-6B jammers and E-2C strike control aircraft.

Things went downhill from there. Two aircraft were shot down, an A-7 and an A-6, and another A-7 was damaged. The A-6 pilot, Lt. Mark Lang, died of his wounds, and his B/N, L. Robert Goodman, was taken prisoner. The downed A-7 pilot was the commander of Carrier Air Wing Six, CDR Edward Andrews. He made it out to sea before ejecting and was picked up by helicopter. The Syrians reported three of their soldiers killed and several injured. The Israelis considered the strike to have been a success based on damage done to artillery and air-defense emplacements.

Nevertheless, the carrier-based Navy was embarrassed by the results and outraged by the Vietnam-era meddling of Washington in planning and tactics at a detailed level. While there wasn't much to be done about civilian leadership, it could do something to insure that future strikes were properly planned and executed. The Naval Strike Warfare Center (NSWC) was established at NAS Fallon, Nevada, in September 1984 to be the counterpart of the famous Top Gun course for fighter pilots. The first class convened in October 1984. More familiarly known as Strike U, various courses cover targeting, intelligence, weaponeering, support assets, threat assessment, and working within the rules of engagement. As one of the last training exercises before deployment, each air wing goes to Fallon with its airplanes to test its readiness. Strike U provides realistic adversary airplanes, simulated air-defense systems, a huge bombing range, and a detailed post-strike critique. A Tactical Aircrew Combat Training System provides for tracking and recording aircraft, actual and simulated weapons, and electronic warfare activity to provide the after-action data for this debrief.

The next strike was against Libya in April 1986 as part of a U.S. campaign to discourage state-sponsored terrorism, specifically the bombing by Libyan agents of a nightclub in West Berlin frequented by U.S. servicemen. This time the strike was well planned and executed at night. The multi-layered and well-equipped Libyan air defense was thoroughly suppressed although one of 18 U.S. Air Force F-111Fs, protected by its own EF-111 jammers, was lost during egress for an undetermined reason. No Navy aircraft were damaged.

In 1996, the Naval Fighter Weapons School (Topgun) and Carrier Airborne Early Warning Weapons School (Topdome) were moved from NAS Miramar, California, and merged with Strike U to form the Naval Strike and Warfare Center (NSAWC). NSAWC is also responsible for the development and evaluation of strike tactics and weapons.

Transition from Deep Strike to Flexible Response

The carrier-based Navy has steadfastly resisted all operations of unpiloted aircraft from aircraft carriers except for the occasional half-hearted operational evaluation. Most weren't particularly effective anyway. However, in the 1980s a suitable deep-strike weapon was qualified, the Tomahawk, that was launched from non-aviation ships. While designated a missile, it had wings and a jet engine like an airplane. Aircraft carrier advocates worried that it would be used to finally justify a reduced number of the big ships or even their elimination. The benefit versus cost of the aircraft carrier was called into question yet again after the first Gulf War in 1991. Aircraft carrier-based strike aircraft had a less publicized role in an Air Force-managed air war that starred the stealthy F-117. The fact that forward-deployed carriers had been in theater less than a week after the invasion of Kuwait, which helped deter the expansion of the Iraqi advance into Saudi Arabia, was as usual not fully appreciated.

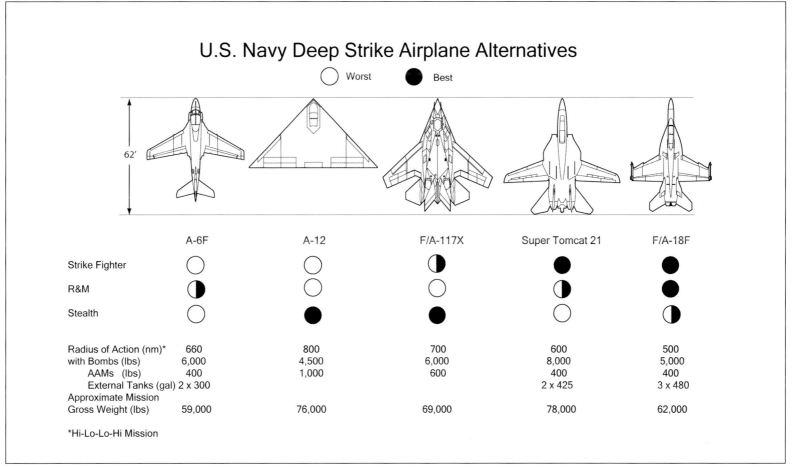

U.S. Navy Deep Strike Airplane Alternatives

○ Worst   ● Best

62'

	A-6F	A-12	F/A-117X	Super Tomcat 21	F/A-18F
Strike Fighter	○	○	◐	●	●
R&M	◑	○	○	◑	●
Stealth	○	●	●	○	◑
Radius of Action (nm)*	660	800	700	600	500
with Bombs (lbs)	6,000	4,500	6,000	8,000	5,000
AAMs (lbs)	400	1,000	600	400	400
External Tanks (gal)	2 x 300			2 x 425	3 x 480
Approximate Mission Gross Weight (lbs)	59,000	76,000	69,000	78,000	62,000

*Hi-Lo-Lo-Hi Mission

The Grumman A-6 Intruder was the carrier-based Navy's main battery for more than three decades. Replacing it was not easy. The A-6F was canceled in favor of the more capable A-12, which was subsequently canceled because it was deemed unaffordable. Both the F/A-117X and Super Tomcat had supporters but, unfortunately, equally powerful critics overly focused on what they perceived as shortcomings. The compromise was the F/A-18F. (Author)

This March 2007 formation flight includes the current U.S. Navy strike-fighter types as well as the U.S. Air Force ones except for the F-16. From the top, they are the F/A-18E, the F-22A, the F-15C, and the F/A-18C. The F-22 was the eventual result of the ATF program, which the Navy decided not to adapt to its requirements. It is inarguably the most capable fighter in the world as well as the most expensive. (In an attempt to improve its perceived cost/benefit ratio by promoting its strike capability, Air Force leadership tried to change its designation to F/A-22 but it didn't stick.) The F-15 is a contemporary of the Navy's F-14, which has been retired. The F-35 will eventually replace the F-15 and the F/A-18C. (U.S. Navy 070316-N-2359T-001)

Boeing (and/or the Navy) was addressing the Super Hornet's capability as a fighter when it loaded this flight test F/A-18E with every air-to-air missile that it could carry. In practice, Hornets carry a mixed load of air-to-air and surface-to-air missiles, although this option would come in handy until air superiority is achieved. (Boeing)

At the time, however, the Navy was well behind the Air Force in employing laser-guided weapons and just had its first stealthy airplane, the A-12, and its premier fighter, the F-14D, shot down by the Secretary of Defense. The Navy quickly recovered by expanding its use of precision-guided weapons, initiating the less technically ambitious but faster-paced F/A-18E/F program in lieu of a marinized F-22 or AF-X, and qualifying the F-14 Bombcat as a placeholder for the Super Hornet's attack capability.[2] As a result, when new crises inevitably erupted and there were urgent calls to action, the carriers and their air wings were always ready to respond. For example, in 2002 following 9/11, carrier-based aircraft including F/A-18Es made many of the initial strikes in Afghanistan because of initial restrictions on operation of Air Force tactical aircraft from neighboring land bases.[3]

It's a tossup as to whether the F-14 fighter pilots or the A-6 attack pilots complained more bitterly about the inferiority of their replacement, the Super Hornet, not to mention the fiscal irresponsibility of spending billions to develop a new airplane when mere hundreds of millions would have sufficed to update and improve both the F-14 and A-6. The counter, simply put, was operational cost savings. The F-14's long-range Phoenix fleet air defense capability was also unlikely to be needed (and had been gradually supplanted by improvements to the Aegis radar/missile combat system based on cruisers and destroyers that had first gone to sea in 1983). The A-6's most demanding payload/range mission had been largely taken over by the Tomahawk.

It wasn't like the admirals had much of a choice in the face of rejections of their plans by OSD and/or Congress. The Boeing (née McDonnell Douglas) F/A-18E/F was essentially all that was left for the Navy to buy until the result of the JSF program became available, if it ever did. The Hornet had prevailed against its many detractors time and again. It survived the charge that the congressional intent was for the Navy to buy a carrier-based version of the Air Force's fighter program; the fact that it was inferior in payload, range, and endurance to the A-7 it was to replace; and its lack of speed and acceleration compared to the F-4. The Super Hornet similarly had to suffer the unfavorable comparison of its outer air battle mission performance compared to the F-14 and the same for its capability as a strike airplane compared to either the A-6 or the F-14, not to mention the F-117.[4]

That's not to say that the Hornet and subsequently the Super Hornet were not the best choices among the alternatives, given the exigencies of the budget versus the requirements. Both airplanes had supporters among the war fighters in the Navy as well as the systems analysts within OSD. The F/A-18s are reliable, maintainable, and have up-to-date avionics. As a fighter, the worse that can be said is that its pilots have to depend on tactics, training, weaponry, and situational awareness for success in air-to-air combat just as their grandfathers did in World War II and their fathers did in Vietnam. For attack pilots, of course, the F/A-18 provided a significant improvement in speed and acceleration.

Although the A-6 bomber had been retired, there was no carrier-based replacement for the essential electronic warfare mission capability of the EA-6B. Nevertheless, its time has also come. After considering a restart of the production line to build a notional EA-6C, the Navy decided in 2002 that it would cost less to develop an electronic attack version of the F/A-18F, the EA-18G, with the added benefit of commonality of basic aircraft systems with the other F/A-18E/Fs in the air wing. The reduction from a minimum of two weapons systems operators to one was a concern, but simulator evaluations of the increased automation of the EWO's tasks allayed it. (The Active Electronically Scanned Array radar being installed in Block 2 production F/A-18E/Fs also provides a self-protection capability that reduces the need for dedicated jamming.)

Boeing received the EA-18G system development and demon-stration (SDD) phase contract in December 2003. Like the EA-6B, the EA-18G is optimized for electronic warfare, with very different avionics and unique appurtenances. The most obvious difference is the addition of large wingtip-mounted pods, which contain the electronic emission receivers. Removable, self-powered pods provide jamming, as on the EA-6B. It can also carry and fire AGM-88 HARM missiles to supplement its jamming capability.

One F/A-18E and two F/A-18Fs were modified to the NEA-18G configuration as part of the development program. The first NEA-18G flew in September 2006. The transition from EA-6Bs in deployable squadrons is planned to begin in 2009 with the Fleet Readiness Squadron, VAQ-129, having received its first airplane in June 2008, even before the start of operational evaluation.

One beneficial outcome is flexibility of asset employment in the carrier air wing. The typical composition changes significantly between Operation Desert Storm in 1991, Operation Iraqi Freedom in 2003, and a notional one within 2010:

In addition to being a fighter, a bomber, and a tanker, the Super Hornet can carry a reconnaissance pod on the centerline sta-tion, seen here on an F/A-18F breaking for a landing on Abraham Lincoln *in June 2008. (The wing pylons are angled outboard for stores separation reasons. The angle of the exter-nal fuel tanks is not a camera lens-induced illusion.)* (U.S. Navy 080612-N-7981E-504)

	1991	2003	2010
F-14	24	10	—
F/A-18	24A/C*	36A+/C/E	50C/E/F
A-6E	10	—	—
EA-6B	4	4	6**
S-3B	10	8	—
E-2C	4	4	4***
Helo	6 H-3	7 SH/HH-60	7 SH/HH-60

* Or A-7E ** Or EA-18G *** Or E-2D

In 1991, the F-14, a significant portion of the air wing's comple-ment of aircraft, had no air-to-ground capability. That resided in the 34 light and medium attack aircraft. Only the A-6E could designate targets for laser-guided bombs. By 2003, there were 46 strike aircraft in the air wing, all of which could designate and deliver guided bombs. In 2010, there would be 50 strike aircraft in the air wing, all of which could also be deployed as fighters if required.

The coated canopies of the EA-6B protect the crew from the high-power transmissions of its jamming pods. The four-place Prowlers along with the E-2C are the last of the Grumman-built airplanes in a carrier's air wing. Soon there will be only the Hawkeye and no Mohicans. (U.S. Navy 041205-N-8704K-005)

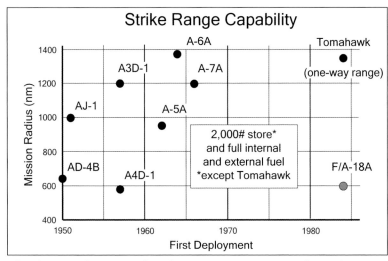

Strike Range Capability

(Chart plotting Mission Radius (nm) vs. First Deployment)

Data points:
- A-6A — ~1380 nm, ~1963
- Tomahawk (one-way range) — ~1380 nm, ~1984
- A3D-1 — 1200 nm, ~1956
- A-7A — 1200 nm, ~1966
- AJ-1 — 1000 nm, ~1950
- A-5A — ~950 nm, ~1962
- AD-4B — ~650 nm, ~1950
- A4D-1 — ~570 nm, ~1956
- F/A-18A — 600 nm, ~1984

2,000# store* and full internal and external fuel *except Tomahawk

The figure of 600 nm for the F/A-18A is not confirmed. It may be closer to 400 to 500.

Like the last year of WWII in the Pacific, if the battle group needs a lot of fighters, all the F/A-18s can be launched as fighters. If air superiority has been assured and ranges are short, then all the F/A-18s can be employed as bombers except for one or two needed as recovery tankers. If the ranges require inflight refueling, any of the F/A-18E/Fs can be configured as a tanker, so the air wing doesn't need to include unique airplanes with limited strike capability for the tanker role. From a maintenance standpoint, only three different airplane types will need to be supported after the EA-18G is introduced, as opposed to five in 1991.

In 2008, the Navy planned to buy about 500 F/A-18E/Fs, half of the originally planned number, and 80 to 90 EA-18Gs. Although the Navy would like to have an all E/F fleet, unless the number of carriers/air wings is reduced, those numbers will result in some squadrons deploying with upgraded Cs for the next decade, just as there were Navy and Marine Corps squadrons flying F/A-18A+s and deploying with air wings in 2008, 20 years after the first strike fighter squadrons deployed with Cs.

The two-seat EA-18G Growler will begin to replace the four-seat EA-6B Prowler as the air wing's electronic reconnaissance and attack aircraft in 2009. This is an F/A-18F configured to represent the EA-18G aerodynamically with dummy antennas, jamming pods, and AGM-88 HARM and AIM-120C air-to-air missiles in addition to two external tanks. (U.S. Navy Photo)

EPILOGUE

Before this artist's concept can be realized, the F-35C will have to successfully stave off arguments that it is too expensive and/or the F/A-18 does not need either replacement or augmentation until an even better alternative is available. (Lockheed Martin *Code One* Magazine)

It is unlikely that the aircraft carrier will disappear from America's arsenal. It is too frequently needed by the nation's leadership to protect or enforce its interests. That means that the aircraft that represent the striking force will continue to have to be upgraded and periodically replaced. Both the F/A-18's replacement and its replacement's prospective replacement/supplement are being developed and defined. The former is likely to be the F-35C, while the latter may possibly be yet another attempt at an unmanned aircraft, this time one that is not only launched from but recovered to the carrier.

Lockheed F-35C

The multi-service, multi-nation F-35 program will result in three similar, 60,000-pound gross weight, single-engine airplanes. All were to have practical stealth incorporated, including internal weapons bays for two 1,000-pound JDAMs and two advanced air-to-air missiles, for first-night strikes against heavily defended targets. The F-35A is a land-based fighter with strike capability. The F-35B is a V/STOL aircraft with the engine-driven counter-rotating fan aft of the cockpit and the swiveling nozzle at the tail. The F-35C is a conventional carrier-based strike fighter. In 2008, the Navy planned to buy 480 to replace the legacy Hornets.

Weapons are continuously being improved, in most cases jointly with the Air Force. For example, the precision achieved, the need to minimize collateral damage, and the desire to maximize the number of targets assignable per aircraft has resulted in a requirement for the unpowered 250-lb Small Diameter Bomb (SDB) with deployable wings for greater standoff. SDB is a joint Air Force/Navy program with the Navy planning to deploy SDB II in 2015. (Boeing)

Coming out of the rising sun, the possible new look of carrier-based aviation: the large, unmanned reconnaissance and strike aircraft. (Northrop Grumman)

One uncertainty for the Navy is when to end the F/A-18E/F production line. As of early 2008, that would be in 2012. However, if the F-35C isn't fielded as planned, or legacy Hornets run out of structural life at a faster rate than planned, more Super Hornets will be required to keep the Air Wings up to strength. The F-35C Critical Design Review was successfully completed in June 2007 and its first flight planned for mid 2009.

However, the basic program has not proceeded according to plan. The first F-35A flew in December 2006, but a redesign was already being accomplished to reduce empty weight and make other essential changes. Moreover, a flight-control system problem in May 2007 resulted in an emergency landing and flight test did not resume until that December. As a result, the first F-35C flight slipped to late 2009. Worse, as of late 2008, the first F-35C deployment was projected as 2015. That opens the door for more F/A-18E/Fs.

Another concern is that the end of Super Hornet production by Boeing means that only one U.S. company, Lockheed Martin, will have a fighter in production. In the never-ending struggle for advantage and market share, Boeing has also lobbied for a program to create a stealthier Super Hornet, "Block 3," in lieu of the F-35 program, allowing time for technology to mature that would allow the development of an even more capable fighter than the F-35, "F/A-XX." Congress, DoD, the Navy, and Lockheed can be sure that whenever F-35 program progress falters or its price increases, as will surely happen from time-to-time, Boeing will have an updated Ultimate Hornet proposition ready for evaluation.

Beyond the F-35C

The Navy is revisiting the World War II concept of a carrier-launched, remotely piloted drone, only now it will be carrier-based with the ability to land back aboard after completing its mission.

Periodically, a visionary in the Navy would postulate and be allowed to develop an unmanned aircraft capability for operation from aircraft carriers. Then, after the briefest of operational evaluations, his superiors would terminate it. World War II spawned the

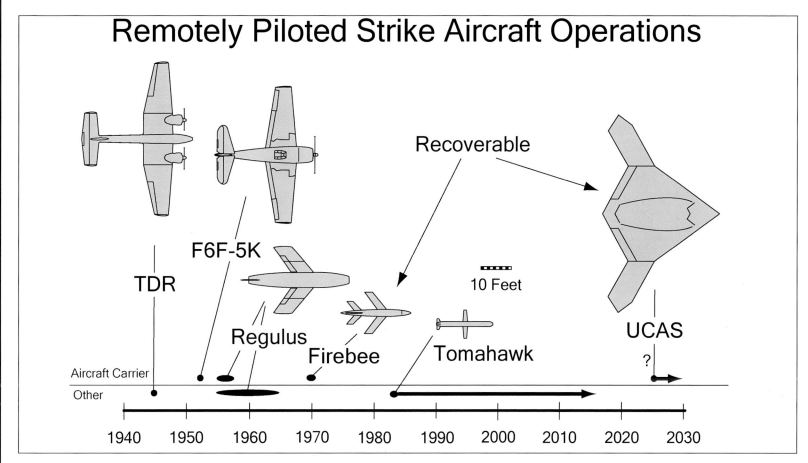

Remotely Piloted Strike Aircraft Operations

Recoverable

10 Feet

TDR

F6F-5K

Regulus

Firebee

Tomahawk

UCAS

?

Aircraft Carrier

Other

1940 1950 1960 1970 1980 1990 2000 2010 2020 2030

The key difference between UCAS and those that have failed to attract support from the aircraft carrier community is its reuse. Shown here as an artist's concept is one about to touch down for an arrested landing. It can then be refueled, rearmed, and relaunched, unlike previous initiatives. (Northrop Grumman)

The operational history of remotely piloted vehicles on aircraft carriers has been brief and never more than experimental, compared to that of the Regulus and Tomahawk on non-aviation ships. The proponents of Unmanned Combat Air System (UCAS), are hopeful that a recoverable, multi-mission remotely piloted aircraft will be more accepted by the carrier Navy. (Author)

TDN/TDR; the Korean War, the F6F-5K; the Cold War, the Regulus; and Vietnam, Firebee for reconnaissance. The size of these aircraft combined with one-time use (except for Firebee) was a significant drawback. The disruption of normal carrier operations when they were launched was also reluctantly tolerated. In contrast, the surface and subsurface Navy embraced Regulus—which led to Polaris in lieu of the A3J—and subsequently Tomahawk, which has taken over a significant portion of the carrier-based Navy's first-day, deep-strike tasking previously accomplished by the A-6. The latest incarnation of unmanned aircraft, UCAS, is more likely to be accepted, if not embraced, by the carrier Navy because it is not single use. One of the selling points that particularly resonates with a Naval aviator is its use as a tanker, an essential capability but not a particularly popular mission to fly. Carried to an extreme if necessary during a mission, the UCAS could be drained of its fuel and sacrificed to save a manned airplane. That and not being capable of stealing your date make it the perfect wingman.

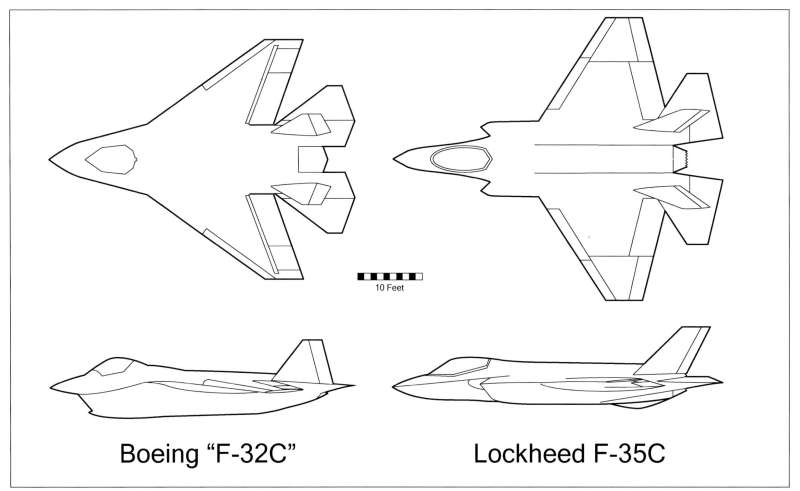

Boeing "F-32C"   ## Lockheed F-35C

10 Feet

Boeing was forced to add a horizontal tail to its JSF proposal for the carrier-based aircraft configuration. The counterpart in the winning Lockheed Martin F-35 proposal features a significantly larger wing than the other versions for the Air Force, U.S. Marine Corps, and foreign air forces. (Author)

In July 2007, Northrop Grumman won a competition to build the X-47B demonstrator of the Unmanned Combat Air System (UCAS). The X-47B is a flying wing, similar to the B-2, with a span of 61 feet and a length of 38 feet, approximately the same size as a strike fighter, with a gross weight of about 45,000 pounds. Payload, in internal bays, is 4,500 pounds with a mission radius of approximately 1,500 nm or a two-hour loiter at 1,000 nm. It is equipped for catapult launch and an arrested landing, with at-sea carrier trials of an unmanned, stealth-compatible aerodynamic shape being the primary purpose of the demonstrator. In March 2008, the Navy was projecting an initial operational capability in 2025 as a complement to an F/A-XX. In October 2008, Northrop Grumman reported that it was on schedule for a first flight planned for November 2009 with carrier trials to begin in November 2011.

The X-47B is the successor to Northrop's much smaller and differently shaped X-47A, which was one of two Defense Advanced Research Projects Agency (DARPA) Unmanned Combat Air Vehicle (UCAV) programs initiated in 2000 to evaluate a carrier-based,

unmanned combat aircraft. The first and only flight of the X-47A was accomplished in February 2003 but it successfully demonstrated a GPS-guided approach to a touchdown 18 feet long and 36 feet to the side of a pre-designated spot. The other contract was with Boeing for the X-46A. However, the DARPA program was terminated in April 2003 in favor of a joint DARPA/USAF/Navy effort before the X-46A had flown or the X-47A could fly again. Boeing had, however, successfully demonstrated the similar X-45A in an earlier DARPA program so they weren't at a disadvantage in the competition. Its first flight was in May 2002. The two X-45As were flown through July 2005, demonstrating formation flight, bomb dropping, mid-mission retargeting, and autonomous attacks.

The joint program, J-UCAS, was subsequently canceled and the Navy continued on its own, more or less at the direction of OSD, in 2006 to develop one for carrier-basing. An operational Navy UCAS would be less observable than a manned aircraft; have two internal weapons bays loaded with 12 precision-guided, 250-pound, small-diameter bombs; and, with inflight refueling, be capable of mission durations measured in days, not hours. Other mission roles include reconnaissance, electronic warfare, and serving as an airborne tanker for other UCAS or manned aircraft.

Cynics might wonder why the Navy plans to take so long to qualify an operational aircraft, given the decades of experience with not only unmanned airplanes but also automatic carrier landings. The first hands-off landings were made at-sea aboard the carrier *Antietam* (CV-36) in August 1957, 50 years ago, by a Douglas F3D Skyknight equipped with the Bell Aerospace Automatic Carrier Landing System. A similar system has been operational for decades for some carrier-based airplanes. An operational flight was made in less than 11 years from the go-ahead for the space shuttle program. Surely an operational carrier-based UCAS could be fielded in less than 18 years, even given the low overall success rate of previous unmanned aerial vehicles (UAV) programs.

Meanwhile, the Air Force has been operating a much smaller remotely piloted, reciprocating-engine-powered drone, the RQ-1 Predator, since 1995. First employed for battlefield reconnaissance in Bosnia, in 2001 it was qualified to fire the laser-guided Hellfire missile for direct interdiction of threats and designated the MQ-1L. With a maximum gross weight of 3,000 pounds, it only carried two small Hellfire missiles. The Predator mission equipment includes a color nose camera for flight control, a satellite link for navigation and two-way communication, a daylight camera, an infrared camera, synthetic aperture radar, and a laser designator. The Hellfire capability was first employed operationally in Yemen in 2001.

An improved and bigger version powered by a turboprop engine, the MQ-9 Reaper, was deployed to Afghanistan in October 2007. With a maximum takeoff gross weight of 10,500 pounds, it has an endurance of 24 hours and can carry four Hellfire missiles and two laser- or GPS-guided bombs. For missions in Iraq and Afghanistan, the Reapers were flown by USAF pilots located at Creech AFB, Nevada, half a world away from the battlefield.

Unmanned aircraft benefits resulting from deletion of an onboard pilot are persistence, expendability, ease of incorporating stealth, indifference to g-loading, and slightly lower empty weight. The main drawbacks of remotely locating the pilot are the reduction in situational awareness, tolerance for ambiguity, and ability to deal with system failures and unanticipated circumstances. Onboard, he or she is also jam proof and not dependent on satellite communication bandwidth for data acquisition and control input.

It seems unlikely that the Navy will ever rely solely on unmanned carrier-based aircraft, if in fact it ever completes the development of one. However, the Air Force's successful employment of unmanned, armed aircraft in combat may foretell a future for the Navy carrier-based strike community of utilizing UCAVs for some missions. It would certainly achieve the ultimate in survivability for their operators as well as capitalize on the hand-eye coordination and cognitive skills of a generation raised on computer games. Until then, however, brave young men—and now women—will continue to be launched in strike aircraft and attack their targets, uncertain of return.

The Air Force already has an operational unmanned and armed aircraft. This MQ-9 Reaper, armed with two of the four Hellfire missiles it can carry, is deployed and being used in combat operations. Like the proposed UCAS, it can be recovered, rearmed, and reflown. (U.S. Air Force 071110-F-1789V-691)

(U.S. Navy 080724-n-7241-002)

RECOMMENDED READING

Books

Barlow, Jeffrey. *Revolt of the Admirals: The Fight for Naval Aviation 1945-1950.* Washington, DC: Ross & Perry, 2001.

Foster, Wynn. *Captain Hook: A Pilot's Tragedy and Triumph in the Vietnam War.* Annapolis, Maryland: Naval Institute Press, 1992.

Hansen, Chuck. *U.S. Nuclear Weapons: The Secret History.* Arlington, Texas: Aerofax Inc., 1988.

Hayward, John T. and C. W. Borklund. *Bluejacket Admiral: The Naval Career of Chick Hayward.* Annapolis, Maryland: Naval Institute Press, 2000.

Heinemann, Edward H. and Rosario Rausa. *Ed Heinemann, Combat Aircraft Designer.* Annapolis, Maryland: Naval Institute Press, 1980.

Holloway, James L. *Aircraft Carriers at War: A Personal Retrospective of Korea, Vietnam, and the Soviet Confrontation.* Annapolis, Maryland: Naval Institute Press, 2007.

Hunt, Peter. *Angles of Attack: An A-6 Intruder Pilot's War.* New York: Ballantine Books, 2002.

Jenkins, Dennis R. *Grumman A-6 Intruder.* North Branch, Minnesota: Specialty Press, 2002.

Kelly, Orr. *Hornet: The Inside Story of the F/A-18.* Novato, California: Presidio Press, 1990.

Levinson, Jeffrey L. *Alpha Strike Vietnam: the Navy's Air War 1964 to 1973.* Novato, California: Presidio Press, 1989.

Miller, Jerry. *Nuclear Weapons and Aircraft Carriers: How the Bomb Saved Naval Aviation.* Washington, DC: Smithsonian Institution Press, 2001.

Morgan, Mark and Rick Morgan. *Intruder: The Operational History of Grumman's A-6.* Atglen, Pennsylvania: Schiffer Military History, 2004.

Piet, Stan and Al Raithel. *Martin P6M Seamaster: The Story of the Most Advanced Seaplane Ever Produced.* Bel Air, Maryland: Martineer Press, 2001.

Rahn, Bob with Zip Rausa. *Tempting Fate: An Experimental Test Pilot's Story.* North Branch, Minnesota: Specialty Press, 1997.

Rausa, Rosario. *Skyraider: The Douglas A-1 "Flying Dump Truck."* Cambridge, England: Nautical & Aviation Publishing Company of America, 1982.

Roland, Buford and William B. Boyd. *U.S. Navy Bureau of Ordnance in World War II.* Washington: U.S. Government Printing Office, 1953.

Skurla, George M. and William H. Gregory. *Inside the Iron Works: How Grumman's Glory Days Faded.* Annapolis, Maryland: Naval Institute Press, 2004.

Stafford, Edward P. *The Big 'E.'* New York: Random House, 1962.

Stevenson, James P. *The Pentagon Paradox: The Development of the F-18 Hornet.* Annapolis, Maryland: Naval Institute Press, 1993.

——*The $5 Billion Misunderstanding: The Collapse of the Navy's A-12 Stealth Bomber Program.* Annapolis, Maryland: Naval Institute Press, 2001.

Stumpf, David K. *Regulus: The Forgotten Weapon System.* Paducah, Kentucky: Turner Publishing Company, 1996.

Thornborough, Anthony M. and Frank B. Mormillo. *Iron Hand: Smashing the Enemy's Air Defences.* Yeovil, England: Haynes Publishing, 2002.

Trimble, William F. *Wings for the Navy: A History of the Naval Aircraft Factory, 1917-1956.* Annapolis, Maryland: Naval Institute Press, 1990.

——*Attack from the Sea: A History of the U.S. Navy's Seaplane Striking Force.* Annapolis, Maryland: Naval Institute Press, 2005.

Zichek, Jared A. *The Incredible Attack Aircraft of the United States: 1948–1949.* Atglen, Pennsylvania: Schiffler Military History, 2009.

Naval Fighters Monographs published by Steve Ginter, Simi Valley, California

Cunningham, Bruce. *Douglas A3D Skywarrior Part One.* Number 45

Cunningham, Bruce and Steve Ginter. *Fleet Whales, Douglas A-3 Skywarrior Part 2.* Number 46

Ginter, Steve. *Douglas A-4A/B Skyhawk in Navy Service.* Number 49

——*Douglas A-4E/F Skyhawk in Navy Service.* Number 51

——*North American A-5/RA-5C Vigilante.* Number 64.

——*North American AJ-1 Savage.* Number 22.

——*North American FJ-4/4B Fury.* Number 25

Koehnen, Rick: *Boeing XF8B-1 Five-in-One Fighter.* Number 65.

Kowalski, Robert. *Grumman AF Guardian.* Number 20.

Kowalski, Bob. *Martin AM-1/1-Q Mauler.* Number 24.

——*Douglas XTB2D-1 Skypirate.* Number 36.

——*Kaiser Fleetwings XBTK-1.* Number 49.

Kowalski, Bob and Steve Ginter. *Douglas XSB2D-1 & BTD-1 Destroyer.* Number 30.

Markgraf, Gerry. *Douglas Skyshark.* Number 43.

Articles

Mares, Ernie LCDR USN (Ret) and David K. Stumpf. "Take Control: Guided Missile Groups One and Two and the Regulus Missile." *The Hook,* Spring 1992, pp 14-39.

Reilly, John C. "Project Yehudi: A Study of Camouflage by Illumination." *American Aviation Historical Society Journal,* Winter 1970, pp 255-262.

Web Sites

ASM-N.2 Bat Guide Bomb
http://biomicro.sdstate/edu/pederses/asmbat.html

George Spangenberg's Oral History
http://www.georgespangenberg.com/gasoralhistory.htm

Special Task Airgroup One—the World War II unit that launched TDR drones against Japanese targets
http://www.stagone.org

GLOSSARY

ADF	Automatic Direction Finding
AESA	Active Electronically Scanned Array (Radar)
AEW	Airborne Early Warning: The detection of enemy air or surface units, usually by radar
AFB	Air Force Base
ARM	Anti-Radiation Missile
ASW	Anti-Submarine Warfare
AP	Armor Piercing
BIS	Board of Inspection and Survey
B/N	Bombardier/Navigator
BOAR	Bureau of Ordnance Aircraft Rocket
BuAer	Bureau of Aeronautics: The U.S. Navy organization responsible for the development and procurement of aircraft between 1921 and 1960
BuWeps	Bureau of Naval Weapons: The U.S. Navy organization responsible for the development and procurement of aircraft and weapons between 1960 and 1966
CAINS	Carrier Airborne Inertial Navigation System
CBO	Congressional Budget Office
CBU	Cluster Bomb Unit: An aircraft store composed of a dispenser and submunitions
CCIP	Continuously Computed Impact Point
CEP	Circular Error Probable: The radius of a circle around the aiming point or MPI within which half of the missiles, bullets, or bombs are expected to or have impacted
Chaff	Metal strips or metal-coated fibers cut to specific lengths to reflect radar emissions as a countermeasure
CNO	Chief of Naval Operations
CTOL	Conventional Takeoff and Land
DIANE	Digital Integrated Attack/Navigation Equipment
DoD	Department of Defense: The Office of the Secretary of Defense, the Military Departments, the Chairman of the Joint Chiefs of Staff, the combatant commands, the Office of the Inspector General of the Department of Defense, the Department of Defense Agencies, field activities, and all other organizational entities in the Department of Defense
Dud	An airplane with a mechanical failure precluding it from being launched
ECM	Electronic Countermeasures
ESM	Electronic Support Measures
EWO	Electronic Warfare Officer: The non-flying systems operator in an electronic warfare aircraft

FIP	Fleet Indoctrination Program
FLIR	Forward-Looking Infrared
FSD	Full Scale Development
GAO	General Accounting Office
Glomb	Glider Bomb
GPS	Global Positioning System: A precision location capability that uses the measurement of range to four or more satellites to establish position. Accuracy is approximately 40 feet, and even better if some of the error can be eliminated by a signal from a nearby ground station of known position
HARM	High-speed Anti-radiation Missile
HUD	Heads-Up Display
HVAR	High-Velocity Aircraft Rocket
IFR	Instrument Flight Rules
INS	Inertial Navigation System
IR	Infrared
JATO	Jet-Assisted Takeoff (In reality, a rocket provided the propulsion, not a jet engine)
JDAM	Joint Direct Attack Munition: An unpowered bomb guided to a designated point by GPS.
JSOW	Joint Standoff Weapon: The AGM-154 is an unpowered winged missile that contains an explosive charge or dispenses area or independently targeted explosives
LABS	Low Altitude Bombing System
LANTIRN	Low Altitude Navigation and Targeting Infrared for Night
LGB	Laser-Guided Bomb
LSO	Landing Signal Officer
Mach	Airspeeds are sometimes referenced to the speed of sound, e.g., Mach 1.0
MER	Multiple Ejector (bomb) Rack
MFD	Multi-Function Display
MPI	Mean Point of Impact: The center of the smallest circle within which half of the missiles, bullets, or bombs have impacted.
NAF	Naval Aircraft Factory
NAS	Naval Air Station
NATC	Naval Air Test Center, NAS Patuxent River, Maryland
NOTS	Naval Ordnance Test Station located at NAS China Lake, California
NWEF	Naval Weapons Evaluation Facility located at Kirtland

	AFB, New Mexico	Shape	A representation of a nuclear weapon used for training
NPE	Navy Preliminary Evaluation	SIOP	Single Integrated Operational Plan: The joint U.S. Air Force/Navy nuclear war operations plan of the United States, options for attacks on a single set of targets listed in the National Target Base.
nm	Nautical Miles		
OpEval	Operational Evaluation	SLAM	Standoff Land Attack Missile
OpNav	Office of the Chief of Naval Operations	SPAD	A nickname for the Douglas AD/A-1 Skyraider
OSD	Office of the Secretary of Defense		
OTS	Over The Shoulder (An atomic bomb delivery maneuver)	TAC	Tactical Air Command
		TACAN	Tactical Air Navigation System
PGM	Precision Guided Missile	TER	Triple Ejector (bomb) Rack
PPI	Plan Position Indicator	TRAM	Target Recognition Attack Multi-sensor
		TRIM	Trails, Roads, Interdiction Multi-sensor
RADAR	RAdio Detection And Ranging		
RATO	Rocket-Assisted Takeoff	UHF	Ultra High Frequency
RCS	Radar Cross Section		
RFP	Request For Proposal	VDI	Vertical Display Indicator
RFQ	Request For Quotation	VFR	Visual Flight Rules
		V/STOL	Vertical/Short Takeoff and Land
SAM	Surface to Air Missile	VTOL	Vertical Takeoff and Land
SAP	Semi-Armor Piercing		
SAR	Search And Rescue	WSO	Weapon Systems Officer: The non-flying systems operator in a combat airplane
SecNav	Secretary of the Navy		
SFC	Specific Fuel Consumption		

END NOTES

Preface

[1] General Dynamics began as the Electric Boat Company. Electric Boat acquired Canadian airplane manufacturer Canadair in 1946. By 1952, aircraft sales were a significant component of the business so a more suitable corporate name was adopted.

[2] For a candid account of Grumman's loss of independency, see *Inside the Iron Works: How Grumman's Glory Days Faded* (Annapolis: Naval Institute Press, 2004) by George M. Skurla and William H. Gregory

[3] The very similar Air Force B-66 retained its designation and the Navy wasn't forced to adopt it.

[4] *Time*, 23 May 1949.

Chapter One: Forged in Battle

[1] The Norden bombsight made famous by the Army Air Forces in World War II was developed by the Navy's Bureau of Ordnance following World War I. Its accuracy was demonstrated in 1931 using the cruiser *Pittsburgh* (CA-4) as a target, albeit a non-maneuvering one.

[2] General purpose (GP) bombs had light casings and half of their weight was explosive, while armor-piercing (AP) or semi-armor piercing (SAP) had lower explosive-to-weight ratios. Armor-piercing bombs had only 15 percent of their weight as explosive, whilst semi-armor piercing bombs had about 30 percent of their weight as explosive.

[3] Torpedoes were very complicated and expensive weapons. They cost approximately $10,000 in 1940, which would be about $150,000 in 2008. Budget limitations resulted in very limited live testing, much less training and practice with an actual torpedo.

[4] Battle of Midway Action Report, Admiral Chester A. Nimitz to Admiral Ernest J. King, CINPac File No. A16 01849, 28 June 1942.

[5] Combat Narrative *The Battle of Midway June 3-6, 1942*, Publication Section, Combat Intelligence Branch, Office of Naval Intelligence, United States Navy, 1943, 58.

[6] "Favorable Reports Continue on the Modified Torpedo," *Naval Aviation Confidential Bulletin*, April 1945, 40.

[7] As reported in *The Hook*, Summer 1985, by Barrett Tillman, Charles Lindbergh made test flights with a bomb-laden Corsair in the Marshall Islands in September 1944. His flights, which each culminated in drops on Japanese installations, began with a single 1,000-pound bomb. He then flew missions with three 1,000-pound bombs, a 2,000-pound bomb, and finally one 2,000 and two 1,000-pound bombs. There is no indication that these overloads became standard practice.

[8] Ironically, the Yagi antenna used for early British and American airborne radar was developed by a Japanese professor but not exploited by the Japanese military until late in the war. In 1926 at Tohoku Imperial University in Japan, Hidetsugu Yagi, in collaboration with Shintaro Uda, created the directional antenna that was originally known as a Yagi-Uda array.

[9] For more details on VT-10's pioneering efforts, see Edward P. Stafford, *The Big 'E.'* (New York: Random House, 1962)

[10] Barrett Tillman, *TBF/TBM Avenger Units of World War 2*, Osprey Publishing, New York, 1999.

[11] The Stearman-Hammond Y-1S, the winner of the Bureau of Air Commerce safe airplane competition, was selected because it had tricycle landing gear, which made it easier to land. The N2C trainers were modified to incorporate a tricycle gear.

[12] RADM D. S. Fahrney, USN (ret) *The History of Pilotless Aircraft and Guided Missiles* (unpublished manuscript, no date), 317.

[13] Interstate proposed a high-speed drone similar to the Northrop flying wing and powered by two Westinghouse 19B jet engines. Maximum speed was projected to be 486 mph at sea level. The XBDR-1 mockup was reviewed in September 1943. BuAer elected not to fund development.

[14] Fahrney, D. S. op cit, 399.

[15] Details of the TDR program can be found at www.stagone.org.

[16] In its first operational use in the Pacific, during a strike on Japanese ships in Balikpapan Harbor, Borneo, one of the Bats homed in on the Pandansari oil refinery instead, which was supposed to be off limits at the request of its Dutch owners. Details of the ASM-N-2 BAT glide bomb program can be found at biomicro.sdstate.edu/pedderses/asmbat.html.

[17] The Brewster SB2A and Vought TBU were initiated at the same time as the SB2C and TBF. The SB2A was initially rated as good as or better than the SB2C. However, Brewster was not able to build an airplane that met the 1940 requirement, much less one with the added armor and self-sealing fuel tanks, or build any at the rate desired. The few that were delivered to the U.S. Navy were relegated to operational training and grounded in 1943 after a spate of tail failures. (The SB2C had teething problems, which meant that the obsolescent SBD did the heavy lifting through the first half of the war.) The TBF was slightly better than the TBU on paper but not as built. However, Vought was unable to capitalize on its advantage due to the emphasis on the Corsair. The project was turned over to Consolidated Vultee, where development and production startup took too long relative to the TBF.

Chapter Two: The Navy's First Attack Aircraft

[1] Air Groups were renamed Air Wings in December 1963. The commander of the air wing continued to be referred to as CAG, however.

[2] The SB2C was eventually qualified to carry a torpedo, but it apparently never did so operationally.

[3] Ed Heinemann, *Ed Heinemann Combat Aircraft Designer*, 104-105.

[4] Other light plane conversions evaluated in the program were the Piper TG-8 as the XLNP-1 and the Aeronca TG-5 as the XLNR-1. Purpose-built gliders were also used, including the Waco CG-4A as the XLRW-1 and the British General Aircraft Ltd Hotspur.

[5] "New Toss Bombing Technique Permits Bomb Release at Longer Ranges and Greater Speeds," *Naval Aviation Confidential Bulletin* dated April 1945, 37-40.

[6] During the AM land-based carrier suitability trials, a violent tail shake was occurring when the tailhook engaged the arresting wire. This culminated with the tail coming off the test airplane. Following extensive testing with a reinforced fuselage and modified tailhook, the AM was finally declared acceptable for carrier landings.

[7] The weight and volume of ASW mission equipment and armament at the time had resulted in two separately configured airplanes operating as a team, one with the big AN/APS-20 radar and the other retaining a bomb bay for torpedoes/depth charges and wing-mounted pylons for rockets.

[8] After some to-ing and fro-ing, with the TB3F-1S/TB3F-2S being designated AF-1/AF-2 for production and then AF-1W/AF-1S, BuAer finally settled on AF-2W/AF-2S in July 1949. S for antisubmarine warfare did not become a primary mission designation until 1950, first used for the Grumman S2F (the SF was a Grumman biplane scout fighter).

[9] Fleet Evaluation of AD-3, -3N, -3Q, and -3W aircraft dated 25 May 1951

[10] The last VA (AW)-35 detachment appears to have become a detachment of VAW-11 before or during the *Lexington* cruise, since its parent squadron was redesignated VA-122 in June 1959 as part of the formation of fleet replacement training squadrons on the West Coast and assigned the responsibility for transitioning pilots into the Skyraider.

Chapter Three: Long Range Nuclear Bombers

[1] Laboratory testing was done to establish critical mass levels. This involved bringing masses of fissile material close together while monitoring the reaction. It was known as "tickling the dragon's tail."

[2] The airplane was named for its usual aircraft commander, Captain Frederick C. Bock, as a play on "boxcar." Some references show it as Bock's Car.

[3] These quotes and other information in this chapter were taken from an article authored by CAPT William E. Scarborough, USN (Ret) in the Fall 1989 issue of *The Hook* magazine.

[4] No carrier landings were attempted, although VADM John T. Hayward told the author in September 1985 that he had made several approaches to, and one touch-and-go on, *FDR* in a P2V. Moreover, he believed that he could have made a full stop landing on the carrier without a hook using the propellers' reverse thrust. However, according to him, qualification of the P2V for arrested landings was not completed because 1) the AJ was coming and 2) it would require above-average piloting skill. Near the end of shore-based qualifications, the test airplane was rendered *hors de combat* by a very hard landing by another pilot, which reinforced Hayward's second point.

[5] William F. Trimble, *Attack from the Sea* (Annapolis, Maryland: Naval Institute Press, 2005): 36-37.

[6] Final Report on Carrier Suitability Tests, Model AJ-1 Airplane, BuNo 122594 dated 11 June 1951: 21.

[7] In 1956, when Morocco achieved independence from France, the town was renamed Kenitra. It is located north of Rabat on the coast of North Africa.

[8] Final Report on Carrier Suitability Tests, Model AJ-1 Airplane, BuNo 122594 dated 11 June 1951: 22.

[9] Of the 55 AJ-1s, at least nine were lost in fatal accidents and as least as many more crashed but with no fatalities. The AJ-2 had a somewhat better record.

Chapter Four: Disappointments

[1] Edward H. Heinemann, *Combat Aircraft Designer* (Annapolis: Naval Institute Press, 1980): 181

[2] Bob Rahn, *Tempting Fate: An Experimental Test Pilot's Story* (North Branch, Minnesota: Specialty Press, 1997): 72.

[3] Much of this A2J history was taken from a BuAer memorandum from the Chief to the Chief of Legislative Liaison dated 17 December 1957 enclosing a report, A2J Data Requested by Senate Preparedness Investigating Subcommittee.

[4] Arthur L. Schoeni was the editor of *Naval Aviation News* in the mid-1940s and early 1950s. He worked in public relations at Vought after leaving the service as a photographer and writer. He continued to write articles on aviation history after retiring from Vought. The *Air Classics* issue was Volume 12, Number 6.

[5] William F. Trimble, *Attack from the Sea* (Annapolis, Maryland: Naval Institute Press, 2005: 97-101.

[6] The complete and well-illustrated story of the P6M SeaMaster program is provided in *Martin P6M SeaMaster* by Stan Piet and Al Raithel.

Chapter Five: The Whale's Tale

[1] Most of the manufacturers who produced carrier-based airplanes for the Navy declined to bid: Chance Vought, Grumman, McDonnell, and North American. I don't know whether this was due to the perceived degree of difficulty of the program, their existing workload, or their lack of familiarity (except for North American) with aircraft in this size class. North American, of course, already had a contract for the A2J.

[2] The proposal design descriptions, overall evaluation, and this quote are from a memorandum to the Chief of the Bureau of Aeronautics dated 21 February 1949, subject: Long Range, Heavy Attack Informal Design Competition—Recommendation, from C. A. Nicholson, Assistant Chief for D&E.

[3] Memorandum from the Chief of Naval Operations to the Chief of the Bureau of Aeronautics, Serial 0205P551 dated 29 April 1949 and signed by J. D. Price, Deputy Chief of Naval Operations (Air).

[4] The escape chute was acceptable for bailout if the A3D was under control but not easily accessed if it was not. The airplane designation was sometimes said to mean "All Three Dead" as a result.

[5] If even more range was required, a fuel tank could be loaded in the bomb bay.

[6] One A3D-1 was re-engined with the Pratt & Whitney J75 and flight tested in 1957 to provide the Navy with flight experience on the engine for the P6M program and Douglas, the DC-8.

[7] Report of Service Acceptance Trials on Model A3D-1 Aircraft dated

7 December 1955.

[8] Service Suitability Trials of Model A3D-1 Airplane, Report #2, Final Report dated 22 October 1956.

[9] John T. Hayward and C. W. Borklund, *Bluejacket Admiral: The Naval Career of Chick Hayward* (Annapolis, Maryland: Naval Institute Press, 2000): 171.

Chapter Six: One Man, One Bomb, One Way?

[1] Another nickname for the LABS maneuver was Harry S, a pun on a scatological term incorporating the first name and middle initial of the only U.S. President to authorize the use of the atomic bomb.

[2] The AD-7 had a slightly different engine and some additional structural beef-up, all transparent to the pilot—the same Flight Manual was used for both airplanes.

[3] Maximum AD endurance was also limited by oil supply. The engine would burn three to four gallons per hour and the oil tank only held 36 gallons.

[4] Rick Morgan, "Where Are They Now? Whitey Feightner," *The Hook* (Fall 2000): 22.

[5] Heinemann's lightweight fighter study was preceded by one for an even more diminutive attack airplane, Design 640, powered by one Westinghouse J34-WE-36 with 3,400-pound thrust. It was intended to carry an Mk 7 and be launched from a submarine using two JATO units.

[6] Bob Rahn, *Tempting Fate: An Experimental Test Pilot's Story* (North Branch, Minnesota: Specialty Press, 1997): pp 138-9.

[7] Letter to the Editor, *The Hook* (Fall 2000): 72.

[8] In June 1956, Grumman proposed an F11F derivative with six stores stations as the A2F for the light attack nuclear-delivery mission. The primary selling point was the benefit of an afterburner in a LABS delivery. The run-in speed was 100 knots faster, there was no buffet in an over-the-shoulder delivery, and, most importantly, the Tiger would be approximately nine miles away when the bomb exploded instead of about five. The Navy chose to continue with the A4D.

[9] A-4D was not used because of the potential for confusion with the original A4D designation.

Chapter Seven: Supersonic Strike

[1] In December 1959, Brig. Gen. Joseph H. Moore set a speed record of 1,216 mph in an F-105 over the 100-km closed course. He reportedly entered the course at over 1,400 mph.

[2] http://www.georgespangenberg.com/

[3] ibid

[4] Report of Nuclear Weapon Aircraft Service Acceptance Trials of Model A-5A (A3J-1) Final Report, U.S. Naval Weapons Evaluation Facility, Kirtland AFB, Albuquerque, New Mexico, NEF 3210 A-5A dated 25 March 1963.

[5] http://www.georgespangenberg.com/

[6] Memorandum from the Deputy Chief of Operations (Air) to the Chief of Naval Operations, Serial 002028P50, dated 11 January 1960.

[7] Fuel tanks in the weapons tunnel were carried over to the reconnaissance derivative. These occasionally broke loose during the catapulting of the aircraft and remained on deck, spilling fuel, which usually caught fire. As least one RA-5C crashed out of control shortly after launch following one of these incidents due to a fire in the tunnel.

Chapter Eight: All-Weather Attack

[1] In the early-1950s, VC-33 evaluated the F3D for the night-attack mission but did not recommend its usage. Compared to the Skyraider, it lacked payload and endurance.

[2] The blue Plexiglas reduced the ambient light in the rear cabin for better readability of the radarscopes.

[3] http://www.georgespangenberg.com/history2.htm

[4] The system was actually more analog in operation than digital, but the computer was digital.

[5] During BIS trials, the A-6 was dived to an indicated Mach Number of 1.025. It had no airspeed red line limitation.

[6] RAdm. Rupe Owens, "Intruder Tactics Mid-60's" *Wings of Gold*, Summer 2004

[7] An A-6E TRAM crew dropped and aimed the laser-guided bomb that blew up the bridge just after "the luckiest man in Iraq" crossed over it in January 1991.

[8] John D. Morrocco, "Navy Considers Phasing Out A-6s," *Aviation Week & Space Technology* (September 27, 1993): 26.

Chapter Nine: Light Attack Redux

[1] The A-7 was one of the first U.S. military aircraft to receive a designation after the change instituted in 1962. It was informally referred to as the A3U at Vought during pre-contract activity.

[2] In the turbofan engine, some of the air from the first few stages of the compression bypassed the combustion process, increasing the mass of air accelerated through the engine. The result compared to an axial-flow engine was more thrust from the same amount of fuel or the same thrust from less fuel. The higher the ratio of bypass air to core flow, the greater the benefit. The drawback was somewhat less thrust at high speeds, initially making it undesirable as a powerplant for supersonic fighters. The first application was therefore on civil jet airliners because of the lower fuel consumption. The Navy was looking for applications for the TF30, which it had developed for the now-canceled F6D Missileer. The A-7 would be one. Another would be the ill-fated F-111B.

[3] These very short response times meant that the contractors had been working on prospective designs for some time.

[4] The qualification of the Bullpup on the F8U-2NE (F-8E) was subsequently canceled, due either to budget considerations or the missile's shortcomings in combat operations. The hump was then used to house ECM avionics.

[5] Although it may not have had any bearing on the competition, it is noteworthy that Ed Heinemann had been forced out of Douglas in 1960. He was their best salesman in addition to being Chief Engineer.

[6] *Reminiscences of Admiral John Smith Thatch*, U.S. Naval Institute Oral history Program, Annapolis, 1977

[7] The maintenance guarantee was also one of the most onerous. If the maintenance requirement exceeded 17 man-hours per flight hour, the

Navy could return the airplane for a complete refund of its cost.

[8] In practice, the P-6 and P-8 were nearly equal in thrust on a hot day.

[9] The only benefit turned out to be a reduction in trim change during flap reduction on takeoff with 30 degrees instead of full flap selected.

[10] Drag counts were associated with external loadings and used to calculate performance from tables. For example, a drag count of 208 represented 20 Mk 82SE bombs on 4 MERs. A drag count of 95 would be six M 117s on the plain pylons.

[11] One wag, intending to disparage both the A-7A's lack of thrust and the F-5's lack of range, noted that the A-7's takeoff distance approximated the F-5's radius of action.

[12] The Air Force used a boom plug-in arrangement because the big strategic bombers needed a higher rate of fuel transfer than could be provided by the probe and drogue system. It also facilitated the hookup of the less maneuverable bomber and the tanker.

[13] Vought was right to be concerned about competition from Douglas and its A-4. The Marines, instead of transitioning to the A-7, purchased the A-4M, with deliveries beginning the same year as the A-7E.

[14] Bob Rahn was only able to meet the original A4D nose wheel liftoff specification by holding the stick full forward during the takeoff roll, compressing the nose wheel strut more and more as his speed increased. When he pulled the stick back at the desired rotation speed, the strut acted like a compressed spring that had been suddenly unloaded, providing the additional nose up moment required. See *Tempting Fate*, page 132.

[15] *Aviation Week*, 17 July 1972, p. 72: The lower cost was undoubtedly still true. In 1973, according to an article in the 6 February issue of *Aviation Week*, the French Navy was reported to have estimated that it could buy only 60 to 80 A-7s for the same cost as 100 A-4s.

Chapter Ten: One if by Land, Two if by Sea

[1] At some point, the F/A-18 was required to have a service life of 6,000 hours compared to the then standard 4,500 hours required of the F-4 and A-7. (Vought claimed that the A-7E had a design life of 8,000 hours.) It was increasingly obvious that airplanes were going to have to last longer due to development and procurement budget constraints. F/A-18As and Cs are being retired at 7.500 hours, impacting Navy planning and budgeting.

[2] Although the reclining position is touted for higher g tolerance, another benefit is lower frontal area, which results in less drag for higher speed and longer range.

[3] According to George Spangenberg, it was "The first formal protest of a Navy source selection in modern history."

[4] Letter from Sol Love, LTV Aerospace Corporation president, to his fellow employees dated 2 June 1975. The F-18 F404 engine was arguably a derivative of the F-17's J101 engine, which was on the Navy's list of engines to be considered.

[5] LTV Aerospace Corporation Press Release dated 5 June 1975.

[6] Ibid.

[7] LTV Aerospace Corporation News Release 75-115 dated 16 July 1975.

[8] http://www.georgespangenberg.com/history3.htm

[9] Lt. Robert E. Stumpf, USN. "One Hornet, Two Stings," Naval Institute *Proceedings*, September 1982, 115-119.

[10] Norman Polmar. "It's a What?," Naval Institute *Proceedings*, January 2003, 105. In DoD 4120.15-L, "Model Designation of Military Aerospace Vehicles," May 12, 2004, it is listed as the FA-18, no slash, which technically would make it an attack airplane with the added capability of a fighter. The lack of a slash, however, is said by some to be the inability of this particular DoD data base to accept one.

[11] Vought almost sold a variant of the A-7X to the Air Force as an upgrade to the A-7D. In 1987, they received a contract for the YA-7F, which was powered by an afterburning Pratt & Whitney F-100. Two were built and flown, the first in late 1989 and the second in early 1990. However, the Air Force not only decided to stick with the A-10 instead, it retired its A-7Ds.

[12] Cycle time refers to the interval between the launch and recovery. Because of time required to reconfigure the flight deck between each launch and recovery, a minimum of 90 minutes flying time was desired, with 105 minutes preferred.

[13] "' Hornet will meet attack spec' says MDC": *Flight International*, 20 November 1982, 1479.

[14] Ibid.

[15] James P. Stevenson. "A Better Hornet Promises, Promises," *Naval Institute Proceedings* (October 1993), 105. The difference in range between the 1981 demonstration, which met the specification, and the production figures is probably due in part to the demonstration not being flown in strict accordance with the specification mission profile, which involves very specific times at throttle settings and altitudes, among other things, that have a significant impact on range.

[16] Benjamin F. Schemmer, "Congress Rejects Land-based Navy Tanker Fleet to Refuel Short-legged F/A-18s," *Armed Forces Journal International*, April 1986, 23.

Chapter Eleven: Replacing the A-6, Round One

[1] The Navy established a program office for NATF at Wright Patterson AFB in September 1988 and contracted for design studies with the manufacturers that received ATF demonstration/validation contracts from the Air Force. However, the Navy's official commitment to NATF ended with the completion of the dem/val phase and so did their involvement, as they chose to proceed with F-14 upgrades instead.

[2] Ironically, some measure of the U.S. advancement in stealth was based on work by a Russian physicist and mathematician, Pyotr Yakovlevich Ufimtsev, who developed equations for predicting the reflection of optical waves from various shapes, now known as the Physical Theory of Diffraction. He was allowed to publish his work internationally in 1962 by the Soviet government since it wasn't considered to be of any military value.

[3] The aerodynamic configuration forced by radar cross section reduction was only flyable if the electronic stabilization and control system was working properly, making its demonstration as important as that of the radar cross section reduction.

[4] In the case of the B-2, the airplane's synthetic aperture radar can be used to precisely locate the target relative to its own GPS-derived position and confer that to the bomb before it is dropped. Most of the GPS signal error was therefore eliminated as it was with Differential GPS.

[5] James P. Stevenson, *The $5 Billion Misunderstanding: The Collapse Of The Navy's A-12 Stealth Bomber Program* (Annapolis: Naval Institute Press, 2001): 48-49.

[6] George M. Skurla, *Inside the Iron Works: How Grumman's Glory Days Faded*

[7] The tabular data is from page 94 of James P Stevenson's *The $5 Billion Misunderstanding: The Collapse of the Navy's A-12 Stealth Bomber Program* (Annapolis, Maryland: Naval Institute Press, 2001).

[8] Ceiling price contracts are a variant of fixed price contracts. Essentially, the contract has a target cost and profit. The contractor and government share the cost or benefit of overruns and underruns, respectively, until the ceiling price is reached, whereupon all additional costs are the responsibility of the contractor.

[9] John D. Morrocco, "Navy Weighs Alternatives After Cheney Kills Avenger 2," *Aviation Week & Space Technology* (January 14, 1991): 18-20.

[10] In 1997, according to GAO report NSIAD-98-152 dated June 1998, the replacement of low observable tape, caulk, paint, and heat tiles on the B-2 due to ordinary wear and tear was a major maintenance burden, contributing to a mission capable rate of only 36 percent, less than half the goal of 77 percent. The processes were time consuming and required an environmentally controlled repair facility. Cure times of some tapes and caulks were as long as 72 hours.

[11] "Dunleavy: Excess Weight Rules Out Use of A-12 Derivative for ATS." *Aerospace Daily* (November 27, 1990): pg 329. ATS was to be a single airframe that would replace the EA-6B for electronic warfare, E-2C for AEW, the S-3 for ASW, and the EX-3 for electronic reconnaissance. ATS never got beyond studies because of budget limitations. The S-3 and ES-3 were retired without replacement. The E-2C and EA-6B, both more essential to the Navy's missions, continue to be upgraded.

Chapter Twelve: Replacing the A-6, Round Two

[1] The GD/McDonnell Douglas/Northrop team was disbanded in early 1993 after Lockheed purchased the General Dynamics Fort Worth facility and products. The Navy would not allow a company to be the prime contractor on two different proposals.

[2] Cheney also cut the B-2 program from 132 aircraft to 21, the C-17 from 210 to 120, and the F-22 from 750 to 648, so it wasn't just a grudge against the Navy. His was also not the final word. Congress continued funding the F-14D for a short while and the C-17 was continued in production for another 20 years.

[3] There was also little or no enthusiasm in the fighter community for the air-to-ground role that the F-4s had to take on early in the Vietnam War.

[4] *Wings of Gold,* Spring 1991, "F-14 Air-to-Ground Program," LCdr Pete Williams, pp 50-51.

[5] Leadership problems in the Navy and at one of its principal subcontractors in the early 1990s affected the Navy's ability to effectively advocate and prosecute its plans. In September 1991, at the annual Tailhook Association convention in Las Vegas, male Naval aviators carried their high-testosterone antics too far even by the low standards of conduct for these affairs. Among other fallouts, some deserved but many not, the Secretary of the Navy was forced to resign. Also during this time, Grumman management alienated both the Navy and important members of Congress to a degree that diminished their ability to influence program decisions in their favor.

[6] An article by Tony Holmes in the Summer 2006 issue of *The Hook* reported that the maintenance manhours for the F-14D during its last cruise were close to 60 manhours per flight hour, whereas the two squadrons of F-18Cs aboard required only 10 to 15. That means the number of maintainers required for a Hornet squadron was one half to two-thirds of a Tomcat squadron's with less "overtime," which can result in errors.

[7] Congressional Record, 30 June 1992, page S9310-12, remarks inserted by Senator Christopher "Kit" Bond who, it should be noted, represented Missouri where McDonnell Douglas was headquartered. For a scathing evaluation of the proposed F/A-18E/G performance, see the remarks of Senator Alfonse D'Amato of New York (Grumman was headquartered on Long Island) on 25 May 1992, page S5931-2.

[8] For one view of Grumman's inability to work with Congress, see *Inside the Iron Works, pg. 170.*

[9] *Aviation Week & Space Technology,* 10 June 1996, "Lantirn Gives Tomcat Night Attack Role," William B. Scott, 40-44.

[10] Patrick J. Finneran Jr. and VADM John Lockhard, USN (Retired). "It's Not Your Father's Hornet," *Proceedings* (October 2001): 80-83.

[11] According to a Rand report, Lessons Learned from the F/A-18 and F/A-18E/F Development Programs dated 2005, the F-22 avionics development cost alone—29 percent of the total development cost—was greater than the total cost for qualification of the initial F-18E/F configuration.

[12] A "snag" in the wing leading edge was a common feature on swept-wing airplanes to create an aerodynamic fence at high angles of attack to minimize span-wise flow that would otherwise occur. It served the same function as a wing fence at less weight and drag. Span-wise flow reduced aileron effectiveness, and at high angles of attack might result in a sudden pitch up due to the wing tip stalling first.

[13] The requirement for the exhaust of the basic engine to be located at the airplane's center of gravity, instead of farther aft as is customary, also dictated a non-optimal configuration for CTOL "F-32s."

Chapter Thirteen: Summary

[1] As described in Chapter 2, an early form of CCIP, the ASG-10, was available in 1945 but apparently not considered reliable and/or useful enough for continued operational use.

[2] The F-22 full-scale development program was initiated in 1991 after a fly-off of the Lockheed and Northrop prototypes. The F/A-18E/F program was approved in 1992. The first F/A-18/Es deployed in July 2002, going straight into combat in Afghanistan. The F-22 was finally declared to be deployable as a 12-plane squadron in December 2005 and flew its first mission in defense of the east coast of the United States (Operation Noble Eagle) in January 2006. It had yet to be used in combat as of early 2009.

[3] As with the first Gulf War, it was a team effort. Tomahawk missiles and long-range Air Force bombers were part of the initial mix. U.S. Air Force and Royal Air Force tanker support was critical because targets in land-locked Afghanistan were up to 900 nm from the carriers.

[4] One wag noted that the increase in the Super Hornet's fuselage length provided the major portion of its increase in range compared to the legacy Hornet. The actual increase in radius was on the order of 100 nm.

INDEX